VLIW Microprocessor Hardware Design

ABOUT THE AUTHOR

Weng Fook Lee is a senior member of the technical staff at Emerald Systems Design Center. He is an acknowledged expert in the field of RTL coding and logic synthesis, with extensive experience in microprocessor design, chipsets, ASIC, and SOC devices. Lee is the inventor/coinventor of 14 design patents and is also the author of *VHDL Coding and Logic Synthesis with Synopsys* and *Verilog Coding for Logic Synthesis.*

VLIW Microprocessor Hardware Design

For ASIC and FPGA

Weng Fook Lee

New York   Chicago   San Francisco   Lisbon   London   Madrid
Mexico City   Milan   New Delhi   San Juan   Seoul
Singapore   Sydney   Toronto

The McGraw·Hill Companies

Library of Congress Cataloging-in-Publication Data

Lee, Weng Fook.
VLIW microprocessor hardware design / Weng Fook Lee.
p.   cm.
ISBN-13: 978-0-07-149702-2
ISBN-10: 0-07-149702-1 (alk. paper)
1. Microprocessors—Design and construction.   I. Title.
TK7895.M5L44   ~~2007~~ 2008
621.39′16—dc22

2007018658

McGraw-Hill books are available at special quantity discounts to use as premiums and sales promotions, or for use in corporate training programs. For more information, please write to the Director of Special Sales, McGraw-Hill Professional, Two Penn Plaza, New York, NY 10121-2298. Or contact your local bookstore.

1 2 3 4 5 6 7 8 9 0   DOC/DOC   0 1 2 1 0 9 8 7

ISBN 978-0-07-149702-2
MHID 0-07-149702-1

This book is printed on acid-free paper.

Sponsoring Editor Stephen S. Chapman	**Proofreader** Shruti Pande
Acquisitions Coordinator Alexis Richard	**Indexer** Steve Ingle
Editorial Supervisor David E. Fogarty	**Production Supervisor** Richard C. Ruzycka
Project Manager Preeti Longia Sinha, International Typesetting and Composition	**Composition** International Typesetting and Composition
Copy Editor Suzanne Lassandro	**Art Director, Cover** Jeff Weeks

Dedicated to my wife,
for all her sacrifices

Contents

Preface

Microcontrollers and microprocessors are used in everyday systems. Basically, any electronic systems that require computation or instruction execution require a microcontroller or microprocessor. Microcontrollers are basically microprocessors coupled with surrounding periphery logic that perform a certain functionality. Therefore, at the core of electronic systems with computational capability (for example, a POS system, an ATM machine, handheld devices, control systems and others) is a microprocessor.

Microprocessors have grown from 8 bits to 16 bits, 32 bits, and currently to 64 bits. Microprocessor architecture has also grown from complex instruction set computing (CISC) based to reduced instruction set computing (RISC) based on a combination of RISC-CISC based and currently very long instruction word (VLIW) based. This book discusses the hardware design and implementation of a 64-bit VLIW microprocessor capable of operating three operations per VLIW instruction word on ASIC and FPGA technology.

The architecture and microarchitecture of the design are discussed in detail in Chapter 2. The ASIC design methodology used for designing the VLIW microprocessor is also discussed by showing each step of the methodology. The VLIW microprocessor begins with the technical specifications which involve the voltage requirements, performance requirements, area utilization, VLIW instruction set, `register file` definition, and details of operation for each instruction. From these technical details, the architecture and microarchitecture consisting of three pipes running in parallel allowing for three operations executed in parallel are described in detail with each pipe being split into four stages of pipelining.

Chapter 3 discusses best known methods (BKM) on RTL coding guidelines which must be met in order to obtain good coding style that can yield optimized synthesis results in terms of area and performance. The reader is shown the importance of each guideline and how it affects the design. Based on these guidelines, the RTL code for each of the modules within the VLIW microprocessor is written. Chapter 3 continues with detailed descriptions of the steps following RTL coding, namely simu-lation, synthesis, standard cell library, layout, DRC, LVS, formal verifi-cation, and physical verification. Creation of testbenches and usage of test plans in verifying the functionality of the RTL code are also discussed. The reader is also shown how code coverage can be used as a method to determine if the testbenches are adequate for verifying the design.

The requirements for synthesis are discussed with topics on stan-dard cell library, design constraints, and synthesis tweaks in Section 3.4. In this chapter, contents and creation of a standard cell library are dis-cussed with information on how the flavors of a standard cell library may affect the synthesis process of a design. For synthesized circuits that cannot meet performance due to tight design constraints, some common methods of design tweaks are discussed.

Section 3.5 shows the reader how formal a verification method can be used to check if a synthesized design matches the golden model of the design (RTL code). If formal verification fails, it indicates that the syn-thesized netlist and the golden RTL code do not match. Formal verifi-cation does not need any stimulus, thereby allowing comparison of the design much quicker compared to gate level simulation.

Section 3.6 discusses pre-layout static timing analysis. During this step of the ASIC flow, the design is checked for setup-time violation and hold-time violation. What these violations are and how they are created are discussed with methods of fixing them.

Section 3.7 addresses the layout portion of the ASIC flow which explains to the readers the three types of layout that can be used for ASIC design, namely custom/manual layout, schematic driven layout, and auto place and route. The advantages and disadvantages of each method are discussed.

Section 3.8 explains what DRC and LVS are, and how they are used to verify the layout of a design. If a design does not pass all the DRC rules, it cannot be sent to a fab for fabrication.

Sections 3.9 to 3.11 describe parasitic extraction and how this infor-mation is back annotated to the design phase to enable an accurate post-layout logic and performance verification of the design. Designs with deep-submicron technologies must always be back annotated to ensure the parasitic does not cause the design to fail. It is common for designs that pass simulation and timing analysis at the pre-layout phase fail at the post-layout phase when parasitic are back annotated.

Section 3.12 describes about tapeout (design completed and ready to be sent to fab). Section 3.13 discusses other issues that need to be con-sidered in design such as clock tree and back annotation.

Section 3.14 shows the reader different methods that are used by designers for low power design. Section 3.15 discusses testability issues. Most designs today are so complex that scan chains are commonly built into the design to allow for ease of testability of the internal logic and external board level connectivity.

Chapter 4 describes a different method to implement the VLIW microprocessor using FPGA. Differences between FPGA and ASIC are explained in this chapter. Advantages and disadvantages of ASIC and FPGA are discussed in detail.

Appendix A shows several examples of testbenches for verifying the functionality and features of the VLIW microprocessor while Appendix B shows the synthesized results and netlist for ASIC and FPGA implementation.

Acknowledgments

This book would not have been possible without the help of many people. I would like to put forward a word of thanks to Prof. Dr. Ali Yeon of the University Malaysia Perlis, Dr. Bala Amawasai, Bernard Lee (CEO of Emerald Systems), Azrul Abdul Halim (Director of Design Engineering, Emerald Systems), Mona Chee, Soo Me, Sun Chong See, Colin Lim, Tim Chen, Azydee Hamid, Steve Chapman, and the staff at McGraw-Hill Professional.

Trademarks

ModelSim, Leonardo Spectrum, FormalPro, IC Station, and IC Station SDL are trademarks of Mentor Graphics Inc.

VCS, Design Compiler, Astro, Formality are trademarks of Synopsys Inc.

NC Verilog, Ambit, Silicon Ensemble, and Incisive are trademarks of Cadence Inc.

Silterra is a trademark of Silterra Malaysia.

VLIW Microprocessor
Hardware Design

Introduction

Microprocessors and microcontrollers are widely used in the world today. They are used in everyday electronic systems, be it systems used in industry or systems used by consumers. Complex electronic systems such as computers, ATM machines, POS systems, financial systems, transaction systems, control systems, and database systems all use some form of microcontroller or microprocessor as the core of their system. Consumer electronic systems such as home security systems, chip-based credit cards, microwave ovens, cars, cell phones, PDAs, refrigerators, and other daily appliances have within the core of their systems either a microcontroller or microprocessor.

What are microcontrollers and microprocessors? If they are such a big part of our daily lives, what exactly are their function?

Microprocessors and microcontrollers are very similar in nature. In fact, from a top level perspective, a microprocessor is the core of a microcontroller. A microcontroller basically consists of a microprocessor as its central processing unit (CPU) with peripheral logic surrounding the microprocessor core. As such it can be viewed that a microprocessor is the building block for a microcontroller (Figure 1.1).

A microcontroller has many uses. It is commonly used to provide a system level solution for things such as controlling a car's electronic system, home security systems, ATM system, communication systems, daily consumer appliances (such as microwave oven, washing machine), and many others.

From a general point of view, a microcontroller is composed of three basic blocks:

1. Memory
   - A nonvolatile memory block to store the program for the microcontroller. When the system is initiated, the microcontroller reads

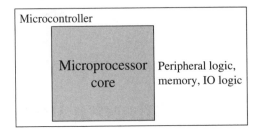

Figure **1.1** Diagram showing microprocessor as core of micro-controller.

the contents of the nonvolatile memory and starts performing its task based on the programming instructions. Examples of nonvolatile memory are electronic programmable read only memory (EPROM), read only memory (ROM), and flash memory.

- A block of volatile memory that is used as temporary storage location by the microcontroller when it is performing its task. When power is turned off from the microcontroller, the contents of the volatile memory are lost. Examples of volatile memory are Random Access Memory (RAM), SRAM, DRAM, DDRRAM, SDRAM, and others.

2. CPU that does all the processing of the instructions read from the nonvolatile memory.

3. Peripheral logic allowing the microcontroller to have access to external IC chips through input/output (IO).

As stated previously, a microprocessor is the CPU of the microcontroller. Within the microprocessor is an arithmetic logic unit (ALU) that allows the microprocessor to process arithmetic and logic instructions provided to the microprocessor.

Our daily lives are filled with use of computers, whether we are aware of it or not. For example, when we go to a bank and make a withdrawal using an ATM, the ATM identifies us and our bank account using an ATM card issued by the bank. That information is relayed from the ATM machine to a central computer system that transmits information back to the ATM regarding the amount of savings in the account and how much can be withdrawn at that moment. When we decide to withdraw a certain sum of money, that transaction is automatically recorded in the bank's central computer system and the corresponding bank account. This process is automated within a computer system, and at the very heart of the computer systems lies many microprocessors.

Computers that we use daily at home or at work have a microprocessor as their brain. The microprocessor does all the necessary functions of the computer when we are using a word editor, spreadsheet, presentation slides, surfing the internet, or playing computer games. Computers cannot function without a microprocessor.

1.1 Types of Microprocessors

The first microprocessor was developed by Intel Corp in 1971. It was called 4004. The 4004 was a simple design compared to the microprocessors that we have today. However, back in 1971 the 4004 was a state-of-the-art microprocessor.

Microprocessors today have grown manifold from their beginnings. Present-day microprocessors typically run in hundreds of megahertz ranging to gigahertz in their clock speeds. They have also grown from 8 bits to 16, 32, and 64 bits. The architecture of a microprocessor has also grown from CISC to RISC and VLIW.

Complex instruction set computing (CISC) is based on the concept of using as little instruction as possible in programming a microprocessor. CISC instruction sets are large with instructions ranging from basic to complex instructions. CISC microprocessors were widely used in the early days of microprocessor history.

Reduced instruction set computing (RISC) microprocessors are very different from CISC microprocessors. RISC uses the concept of keeping the instruction set as simple as possible to allow the microprocessor's program to be written using only simple instructions. This idea was presented by John Cocke from IBM Research when he noticed that most complex instructions in the CISC instruction set were seldom used while the basic instructions were heavily utilized.

Apart from the CISC and RISC microprocessors, there is a different generation of microprocessor based on a concept called very long instruction word (VLIW). VLIW microprocessors make use of a concept of instruction level parallelism (ILP)—executing multiple instructions in parallel.

VLIW microprocessors are not the only type of microprocessors that take advantage of executing multiple instructions in parallel. Superscalar superpipeline CISC/RISC microprocessors are also able to achieve parallel execution of instructions.

1.2 Types of Microprocessor Architecture

To achieve high performance for microprocessors, the concept of pipeline is introduced into microprocessor architecture. In pipelining, a microprocessor is divided into multiple pipe stages. Each pipe stage can execute an instruction simultaneously. When a stage in the pipe has completed executing its instruction, it will pass the results to the next stage for further processing while it takes another instruction from its

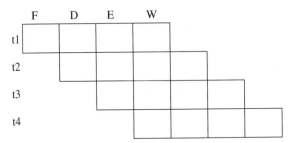

Figure 1.2 Diagram showing instruction execution for pipeline microprocessor.

preceding stage. Figure 1.2 shows the instruction execution for a pipeline microprocessor that has the four basic stages of pipe:

1. fetch—This stage of the pipeline fetches instruction/data from instruction cache/memory.

2. decode—This stage of the pipeline decodes the instruction fetched by the fetch stage. The decode stage also fetches register data from the register file.

3. execute—This stage of the pipeline executes the instruction. This is the stage where the ALU (arithmetic logic unit) is located.

4. writeback—This stage of the pipeline writes data into the register file.

A pipeline microprocessor as shown in Figure 1.2 consists of basic four stages. These stages can be further subdivided into more stages to form

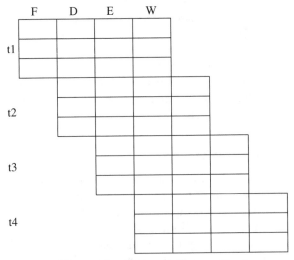

Figure 1.3 Diagram showing instruction execution for superscalar pipeline microprocessor.

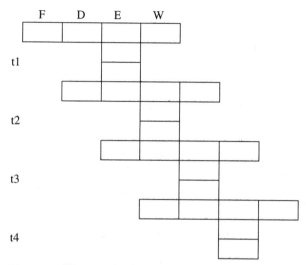

Figure 1.4   Diagram showing instruction execution for VLIW microprocessor.

a superpipeline microprocessor. A superpipeline microprocessor has the disadvantage of requiring more clock cycles to recover from a branch instruction compared to a fewer-stage pipeline microprocessor.

To achieve multiple instruction execution, multiple pipes can be put together to form a superscalar microprocessor. A superscalar microprocessor increases in complexity but allows multiple instructions to be executed in parallel. Figure 1.3 shows the instruction execution for a superscalar pipeline microprocessor.

VLIW microprocessors use a long instruction word that is a combination of several operations combined into one single long instruction word. This allows a VLIW microprocessor to execute multiple operations in parallel. Figure 1.4 shows the instruction execution for a VLIW microprocessor.

Although both superscalar pipeline and VLIW microprocessors can execute multiple instructions in parallel, each microprocessor is very different and has its own set of advantages and disadvantages.

Superscalar pipeline	VLIW
Multiple instructions issued per cycle	One VLIW word is executed per cycle. However each VLIW word consists of several instructions.
Hardware is complex as the microprocessor has multiple instructions incoming.	Hardware is simpler as the microprocessor has a single VLIW word incoming.
Compiler is not as complicated as that of VLIW compiler.	Compiler is complicated as the compiler needs to keep track of the scheduling of instructions.
Smaller program memory is needed.	Larger program memory is needed.

VLIW microprocessors typically require a compiler that is more complicated as it needs to ensure that code dependency in its long instruction word is kept to a minimum.

1. The VLIW microprocessor takes advantage of the parallelism achieved by packing several instructions into a single VLIW word and executing each instruction within the VLIW word in parallel. However, these instructions must have dependency among them kept to a minimum, otherwise the VLIW microprocessor would not be efficient. VLIW microprocessors rely heavily on the compiler to ensure that the instructions packed into a VLIW word have minimal dependency. Creating that "intelligence" into a VLIW compiler is not trivial; much research has been done in this area. This book does not discuss how an efficient compiler can be created or compiler concepts for VLIW, but concentrates instead on the hardware design of a VLIW microprocessor and how it can be achieved using Verilog HDL.

2. VLIW uses multiple operations in a single long instruction word. If one operation is dependent on another operation within the same VLIW word, the second operation may have to wait for the first operation to complete. In these situations, the compiler would insert NOP (no operation) into the VLIW word, thereby slowing down the efficiency of the VLIW microprocessor. To look at the problem of operation dependency during the execution of the operation, let us assume a VLIW instruction that consists of two operations.

```
add r0, r1, r2 : add r2, r3, r4
```

Because the operations in the VLIW instruction are dependent (second operation of add r2, r3, r4 needs the result from the first operation, add r0, r1, r2), the second operation cannot execute until the first operation is complete. The simplest solution would be for the compiler to insert NOP between the two operations to ensure that the results of the first operation are ready when the second operation is executed. VLIW instruction after insertion of NOP:

```
add r0, r1, r2 : NOP
```

```
NOP : add r2, r3, r4
```

As a result of the NOP insertion, there will be two VLIW instructions instead of one. Assuming that the VLIW microprocessor is a four stage pipeline with the first stage fetch, second stage decode, third stage execute, and final stage writeback (each stage of the VLIW microprocessor is explained in detail in Section 2.3), the VLIW instructions which consist of two operations per instruction will enter the pipeline serially.

add r0,r1,r2				Time T1
NOP				
NOP	add r0,r1,r2			Time T2
add r2,r3,r4	NOP			
	NOP	add r0,r1,r2		Time T3
	add r2,r3,r4	NOP		
		NOP	add r0,r1,r2	Time T4
		add r2,r3,r4	NOP	

By inserting NOP into the VLIW instruction, the first operation of add r0, r1, r2 is executed by the microprocessor before the second operation of add r2, r3, r4 is executed. This ensures that the issue of operation dependency is avoided. However, the disadvantage of this method is that the instruction code size will increase while performance of the microprocessor is affected. There are two possible solutions to this problem:

1. The VLIW compiler ensures that there is no dependency between operations within a VLIW instruction.
2. Implement hardware register bypass logic between operations of a VLIW instruction. Register bypass implementation is discussed in Section 3.2.4.

Design Methodology

To design a VLIW microprocessor, the first step is to determine the design methodology. The methodology will show each step that needs to be taken from the beginning of the microprocessor design to verification and final testing. Figure 2.1 shows the design methodology that is used for the design of the VLIW microprocessor.

2.1 Technical Specification

This is the beginning of the design methodology flow. In this step, the technical features and capability of the VLIW and superscalar pipeline microprocessor are defined. The specifications will influence the architecture and microarchitecture of the microprocessor. From the specifications, all design considerations are made with respect to meeting the specified technical requirements. A list of the technical specifications for the design and implementation of the microprocessor follows:

- Must be able to operate at 3.0V conditions
    - In order for the design to operate in 3.0V conditions, the fab process technology considered for doing the design must be able to support 3.0V operation.
    - Normally, the chosen fab for fabricating the design will have different technology catered to different operating voltages and design requirements. The technologies provided by the fab may cover 5V operations, 3V operations, 1.8V operations or lower, mixed signal design, logic design, or RF design.

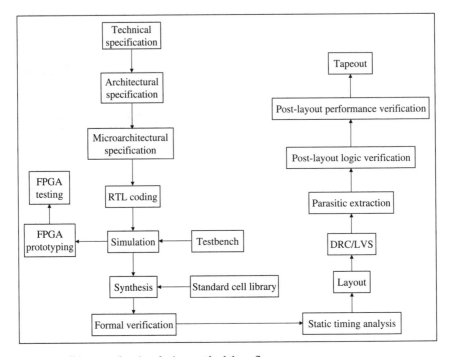

Figure 2.1   Diagram showing design methodology flow.

- Performance must meet a minimum of 200 MIPS (200 million instructions per second)
  - □ This is an important requirement that will have great impact on the architectural specification. By having a minimum requirement of 200 MIPS, the architecture of the VLIW microprocessor must be able to operate under conditions that can achieve such speed. For example, if the microprocessor can operate at 100 MHz, it must execute two instructions at any one time in order to achieve 200 MIPS.
- Microprocessor operates in 64 bits.
  - □ Data bus and internal registers must be architectured to 64 bits.
- Area of design implementation must be kept to a minimum to reduce cost. Transistor count should not exceed 400,000 to limit the size.
- The microprocessor has sixteen internal registers, 64 bits each. This will form the `register file` of the microprocessor. Each register is addressed using 5 bits, ranging from address R0000 for register 0 to address R1111 for register 15. The most significant bit of the address is reserved for future expansion (see Table 2.1).

Note: VLIW microprocessors commonly have larger register file. It is common for VLIW microprocessors to have 256 registers or more. Having a large register file allows the microprocessor to store more data internally rather than externally. This boosts performance as access to register file is faster compared to external memory access. However, having a large register file increases the die size. A balance needs to be achieved between register file size and die size. For ease of understanding, the VLIW microprocessor example is defined with only 16 registers.

■ Instruction sets include arithmetic operations, load operations, read of internal register operations, and compare operations. Section 2.1.1 explains the operations defined for the microprocessor.

2.1.1   Instruction Set of VLIW Microprocessor

When creating the instruction set for the VLIW microprocessor, the operations of arithmetic, load, read, and compare are considered and included in the instruction set. For the design of the VLIW microprocessor, an arithmetic and logic set of 16 operations is created. The list of operations is shown in Table 2.2, with each operation represented by a 5-bit code, with the most significant bit being a reserved bit for future operation code expansion.

TABLE 2.1   Register Address for Internal Register of register file

Internal Register Name	Register Address
r0	R0000
r1	R0001
r2	R0010
r3	R0011
r4	R0100
r5	R0101
r6	R0110
r7	R0111
r8	R1000
r9	R1001
r10	R1010
r11	R1011
r12	R1100
r13	R1101
r14	R1110
r15	R1111

> *Note*: VLIW microprocessors commonly have anywhere up to 64 instructions or more. For ease of understanding, the VLIW microprocessor example is defined with only 16 instructions.

TABLE 2.2   Operation Code for the VLIW Microprocessor Instruction Set

Operation	Code
nop	R0000
add	R0001
sub	R0010
mul	R0011
load	R0100
move	R0101
read	R0110
compare	R0111
xor	R1000
nand	R1001
nor	R1010
not	R1011
shift left	R1100
shift right	R1101
barrel shift left	R1110
barrel shift right	R1111

Each operation code is combined with the internal register address to form an arithmetic or logic operation. Each operation consists of 5 bits for defining the operation code (as shown in Table 2.2) and 15 bits for defining the internal register addresses (as shown in Table 2.1). In total, an operation will consists of 20 bits. Table 2.3 shows how the different bits of the operation code and internal register addresses are combined to form an operation.

TABLE 2.3   Combination of Operation Code and Internal Register Addresses to Form an Operation

Bits [19:15]	Bits [14:10]	Bits [9:5]	Bits [4:0]
Operation code	source1 address	source2 address	destination address

The columns for source1, source2 and destination address are internal register addresses. The VLIW microprocessor has sixteen internal registers and each is defined with its own register address as shown in Table 2.1.

Section 2.1.2 explains how each operation code can be used with the internal register addresses to form an operation.

2.1.2 Definition of Opcode for VLIW Instruction Set

The operation code shown in Table 2.2 consists of 5 bits, with the most significant bit being a reserved bit for future expansion. Bits 3 to 0 are used to represent the 16 different possible operations. Similarly, each internal register is assigned five address bits with the most significant bit being a reserved bit for future expansion, as shown in Table 2.1.

1. Operation code R0000—nop

   This operation code is for a "no operation" performed. This means that the VLIW microprocessor is idle when this operation code is decoded. Table 2.4 shows the bit format for operation code nop. Bits 19, 14, 9, and 4 are reserved bits. For the nop, the internal register addresses of source1, source2, and destination are ignored because no internal register access is required.

TABLE 2.4 Bit Format for Operation Code nop

Bits [19:15]	Bits [14:10]	Bits [9:5]	Bits [4:0]
R0000	RXXXX	RXXXX	RXXXX

2. Operation code R0001—add

   This operation code is for arithmetic addition. The VLIW microprocessor will perform an addition of data from internal registers specified by source1 and source2, and write the results of the addition into the internal register specified by destination.

   destination = source1 + source2

   Since all the internal registers are 64 bits, if an addition creates a result that have a carry out, it is ignored. Only the sum of the addition is written into the 64 bit destination register (shown in Table 2.5).

TABLE 2.5 Bit Format for Operation Code add

Bits [19:15]	Bits [14:10]	Bits [9:5]	Bits [4:0]
R0001	source1	source2	destination

3. Operation code R0010—sub

   This operation code is for arithmetic subtraction. The VLIW microprocessor will perform a subtraction of data from internal registers specified by source2 from source1, and write the results of the subtraction into the internal register specified by destination.

   destination = source1 - source2

If the results of the subtraction creates a borrow, it is ignored (shown in Table 2.6).

TABLE 2.6   Bit Format for Operation Code sub

Bits [19:15]	Bits [14:10]	Bits [9:5]	Bits [4:0]
R0010	source1	source2	destination

4. Operation code R0011—mul

   This operation code is for arithmetic multiplication. The VLIW microprocessor will perform a multiplication of data from internal registers specified by source1 and source2, and write the results of the multiplication into the internal register specified by destination.

   $$destination = source1 * source2$$

   For the multiply operation code, the data at source1 and source2 are limited to the lower 32 bits even though the internal registers source1 and source2 are 64 bits. The results of the multiply operation is 64 bits (shown in Table 2.7).

TABLE 2.7   Bit Format for Operation Code mul

Bits [19:15]	Bits [14:10]	Bits [9:5]	Bits [4:0]
R0011	source1 (limited to lower 32 bit contents)	source2 (limited to lower 32 bit contents)	destination

5. Operation code R0100—load
   This operation code is for loading data into an internal register. The VLIW microprocessor will load the data from the 64 bit data bus input into an internal register specified by destination (shown in Table 2.8).

   $$destination = data\ on\ data\ bus$$

TABLE 2.8   Bit Format for Operation Code load

Bits [19:15]	Bits [14:10]	Bits [9:5]	Bits [4:0]
R0100	RXXXX	RXXXX	destination

6. Operation code R0101—move
   This operation code is for moving data from one internal register to another. The VLIW microprocessor will move the contents of the

internal register specified by source1 to the internal register specified by destination (shown in Table 2.9).

$$destination = source1$$

TABLE 2.9   Bit Format for Operation Code move

Bits [19:15]	Bits [14:10]	Bits [9:5]	Bits [4:0]
R0101	source1	RXXXX	destination

7. Operation code R0110—read
   This operation code is for reading of data from an internal register. The VLIW microprocessor will read the contents of internal register specified by source1 and send the data to the output port of the microprocessor (shown in Table 2.10).

   $$Output\ port = source1$$

TABLE 2.10   Bit Format for Operation Code read

Bits [19:15]	Bits [14:10]	Bits [9:5]	Bits [4:0]
R0110	source1	RXXXX	RXXXX

8. Operation code R0111—compare
   This operation code is for arithmetic comparison. The VLIW microprocessor will perform a comparison of data from internal registers specified by source1 and source2 and the outcome of the comparison will set the appropriate bit of the internal register specified by destination (shown in Table 2.11).

   If the data of source1 is compared equal to the data of source2, a jump is executed (branch to another instruction).

TABLE 2.11   Bit Format for Operation Code compare

Bits [19:15]	Bits [14:10]	Bits [9:5]	Bits [4:0]
R0111	source1	source2	destination

   i. source1 = source2   → Branch to another instruction, a jump is required
  ii. source1 > source2   → Bit 1 of destination register = 1
 iii. source1 <= source2   → Bit 2 of destination register = 1
 iv. source1 <= source2   → Bit 3 of destination register = 1
  v. source1 >= source2   → Bit 4 of destination register = 1
 vi. All other bits of destination register are set to 0.

9. Operation code R1000—xor
   This operation code is for XOR function. The VLIW microprocessor perform an XOR function on data from internal registers specified by source1 and source2, and write the results into the internal register specified by destination (shown in Table 2.12).

TABLE 2.12   Bit Format for Operation Code xor

Bits [19:15]	Bits [14:10]	Bits [9:5]	Bits [4:0]
R1000	source1	source2	destination

10. Operation code R1001—nand
    This operation code is for NAND function. The VLIW microprocessor will perform a NAND function on data from internal registers specified by source1 and source2, and write the results into the internal register specified by destination (shown in Table 2.13).

TABLE 2.13   Bit Format for Operation Code nand

Bits [19:15]	Bits [14:10]	Bits [9:5]	Bits [4:0]
R1001	source1	source2	destination

11. Operation code R1010—nor
    This operation code is for NOR function. The VLIW microprocessor will perform a NOR function on data from internal registers specified by source1 and source2, and write the results into the internal register specified by destination (shown in Table 2.14).

TABLE 2.14   Bit Format for Operation Code nor

Bits [19:15]	Bits [14:10]	Bits [9:5]	Bits [4:0]
R1010	source1	source2	destination

12. Operation code R1011—not
    This operation code is for NOT function. The VLIW microprocessor will perform a NOT function on data from internal register specified by source1 and write the results into the internal register specified by destination (shown in Table 2.15).

TABLE 2.15   Bit Format for Operation Code not

Bits [19:15]	Bits [14:10]	Bits [9:5]	Bits [4:0]
R1011	source1	RXXXX	destination

13. Operation code R1100—shift left

This operation code is for shifting left function. The VLIW micro-processor will perform a shift left function on data from internal registers specified by source1 and write the results in internal register specified by destination. The amount of bits that shift left on source1 is decoded by bits [3:0] of source2. For example, if source2 [3:0] is 0001, source1 is shifted left by one bit. If source2 [3:0] is 1001, source1 is shifted left by nine bits. When shifting left, the least significant bit appended to source1 is logic zero. Since only bits [3:0] of source2 is decoded, the shift left operation code can only shift left a maximum of 15 bits at any one time (shown in Table 2.16).

TABLE 2.16   Bit Format for Operation Code shift left

Bits [19:15]	Bits [14:10]	Bits [9:5]	Bits [4:0]
R1100	source1	source2	destination

14. Operation code R1101—shift right

This operation code is for shifting right function. The VLIW micro-processor will perform a shift right function on data from internal registers specified by source1 and write the results in internal register specified by destination. The amount of bits that shifts right on source1 is decoded by bits [3:0] of source2. For example, if source2 [3:0] is 0001, source1 is shifted right by 1 bit. If source2 [3:0] is 1001, source1 is shifted right by 9 bits. When shifting right, the most significant bit appended to source1 is a zero. Because only bits [3:0] of source2 are decoded, the shift right operation code can only shift right a maximum of 15 bits at any one time (shown in Table 2.17).

TABLE 2.17   Bit Format for Operation Code shift right

Bits [19:15]	Bits [14:10]	Bits [9:5]	Bits [4:0]
R1101	source1	source2	destination

15. Operation code R1110—barrel shift left

This operation code is for barrel shift left function. The VLIW micro-processor will perform a barrel shift left function on data from internal registers specified by source1 and write the results in internal register specified by destination. The amount of bits that barrel shift left on source1 is decoded by bits [3:0] of source2. For example, if source2 [3:0] is 0001, source1 is barrel shifted left by 1 bit. If source2 [3:0] is 1001, source1 is barrel shifted left by 9 bits. When barrel shifting left, the most significant bit becomes the least significant bit. Because only bits [3:0] of source2 are decoded, the barrel shift left operation code can only barrel shift left a maximum of 15 bits at any one time (shown in Table 2.18).

TABLE 2.18   Bit format for Operation Code `barrel shift left`

Bits [19:15]	Bits [14:10]	Bits [9:5]	Bits [4:0]
R1110	source1	source2	destination

16. Operation code R1111—`barrel shift right`
    This operation code is for barrel shift right function. The VLIW microprocessor will perform a barrel shift right function on data from internal registers specified by `source1` and write the results in internal register specified by `destination`. The amount of bits that barrel shift right on `source1` is decoded by bits [3:0] of *source2*. For example, if `source2` [3:0] is 0001, `source1` is barrel shifted right by 1 bit. If `source2` [3:0] is 1001, `source1` is barrel shifted right by 9 bits. When barrel shifting right, the least significant bit becomes the most significant bit. Because only bits [3:0] of `source2` are decoded, the barrel shift right operation code can only barrel shift right a maximum of 15 bits at any one time (shown in Table 2.19).

TABLE 2.19   Bit Format for Operation Code `barrel shift right`

Bits [19:15]	Bits [14:10]	Bits [9:5]	Bits [4:0]
R1110	source1	source2	destination

2.1.3   Definition of VLIW Instruction

Section 2.1.2 describes the definition of the operation code for the VLIW microprocessor. These operation codes are combined together to form a V̲ery L̲ong I̲nstruction W̲ord. Each VLIW instruction word is 64 bits and consists of three operations. Each of the operations can be any one of the operation codes described in Section 2.1.2. For example, let us assume that there are three operations (`add, sub, move`) combined to form a VLIW instruction word.

Note: VLIW microprocessors commonly have between 64 and 1024 bits for a VLIW instruction word, while some have variable length. For ease of understanding, the VLIW microprocessor example is defined with 64-bit VLIW instruction word.

Operation 1: add r0,r1,r2

Bits [19:15]	Bits [14:10]	Bits [9:5]	Bits [4:0]
R0001	R0000	R0001	R0010

Operation 2: sub r3,r4,r5

Bits [19:15]	Bits [14:10]	Bits [9:5]	Bits [4:0]
R0010	R0011	R0100	R0101

Operation 3: move r10, r8

Bits [19:15]	Bits [14:10]	Bits [9:5]	Bits [4:0]
R0101	R1010	RXXXX	R1000

VLIW instruction word:

Bits [64:60]	Bits [59:40]	Bits [39:20]	Bits [19:0]
RRRR	R0001R0000R0001-R0010 add r0, r1, r2 Operation 1	R0010R0011R0100-R0101 sub r3, r4, r5 Operation 2	R0101R1010-RXXXXR1000 move r10, r8 Operation 3

R indicates the reserved bits for future expansion; X indicates don't care for the corresponding operation.

The three operations combined to form one VLIW instruction word allows the VLIW microprocessor to read one instruction but execute three operations in parallel.

2.2   Architectural Specification

Section 2.1 describes the technical specification for the VLIW microprocessor. From the technical specification, the architectural specification is derived. This is a crucial step because the architecture of a design plays an important part in the performance capability and area utilization of the design. For example, if the microprocessor is architectured for 100 MHz, designing it to achieve performance greater than 100 MHz will be a difficult task. The architecture of a design plays a major role in the overall capability of a design.

Figure 2.2 shows a generic architecture that can be used to represent the VLIW microprocessor. The microprocessor fetches instructions from an external instruction cache into its internal instruction buffers and decoders. The instruction is then passed on to multiple execution units which allows for multiple operations to be executed in parallel.

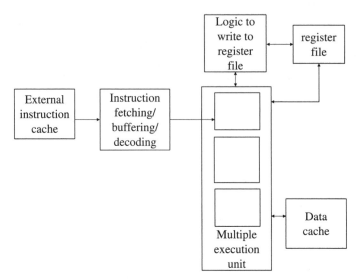

Figure 2.2 Diagram showing a generic architecture for VLIW microprocessor.

Based on the technical specifications described in Section 2.1 and the generic architecture diagram of Figure 2.2, the VLIW microprocessor can be simplified and architectured using a pipeline technology of four stages:

1. The VLIW microprocessor is architectured to take advantage of the pipeline technology. (For further information on pipeline technology, please refer to Hennessy and Patterson, *Computer Architecture: A Quantitative Approach* [Morgan Kaufmann], and Patterson and Hennessy, *Computer Organization & Design: The Hardware / Software Interface* [Morgan Kaufman].)

2. Each 64-bit VLIW instruction word consists of three operations. To maximize the performance capability, the architecture is built to execute the three operations in parallel. Each operation is numbered and categorized as pipe1, pipe2, and pipe3 with pipe1 operating operation 1, pipe2 operating operation 2 and pipe3 operating operation 3.

3. Each operation is split into four stages: fetch stage, decode stage, execute stage, and writeback stage. Four stages are chosen to keep the architecture simple yet efficient. The fetch stage fetches the VLIW instruction and data from external devices such as memory. The decode stage decodes the VLIW instruction to determine what operations each pipe needs to execute. The execute stage executes the operation decoded by the decode stage. The writeback stage (the last stage of the pipe) writes the results from the execution of the instruction into internal registers.

4. All three operations share a set of sixteen 64-bit internal registers, which forms a `register file`. During the `decode` stage, data are read from the `register file` and during `writeback` stage, data are written into the `register file`.

Based on these requirements, the VLIW microprocessor is architectured to Figure 2.3.

In Figure 2.3, the incoming instructions and data from external systems to the VLIW microprocessor are fetched by the `fetch` unit, the first stage of the VLIW microprocessor.

After the instruction and data have been fetched, it is passed to the `decode` stage. The 64-bit instruction consists of three operations (refer Section 2.1.3). Each operation is passed to the corresponding `decode` stage. Each operation is also passed from the `fetch` stage to the `register file` to allow the data to be read from the `register file` for each corresponding operation.

In the `decode` stage, the operations are decoded and passed onto the `execute` stage. The `execute` stage, as its name implies, will execute the corresponding decoded operation. The `execute` stage has access to the shared `register file` for reading of data during execution.

Upon completion of execution of an operation, the final stage (`writeback` stage) will write the results of the operation into the `register file`, or read data to the output of the VLIW microprocessor for `read` operation.

Table 2.20 describes the interface signals defined for the architecture of the VLIW microprocessor. Figure 2.4 shows the interface signal diagram of the VLIW microprocessor.

To allow ease of understanding on the implemented RTL code of the VLIW microprocessor, the following are taken into consideration:

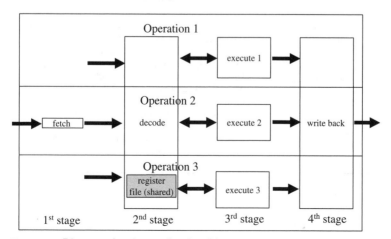

Figure 2.3 Diagram showing top level architecture.

TABLE 2.20   Description of VLIW Microprocessor Interface Signals

Pin Name	Input/ Output	Bit Size	Description
clock	input	1	Input clock pin. The VLIW microprocessor is active on rising edge clock.
reset	input	1	Input reset pin. Reset is asynchronous and active high.
word	input	64	The 64-bit word represents the VLIW instruction from external instruction memory. The 64 bits are represented as: ▪ Bits 64 to 60, 59, 54, 49, 44, 39, 34, 29, 24, 19, 14, 9, and 4 are reserved bits. ▪ Bits 58 to 55 represent opcode for operation 1. ▪ Bits 53 to 50 represent source1 for operation 1. ▪ Bits 48 to 45 represent source2 for operation 1. ▪ Bits 43 to 40 represent destination for operation 1. ▪ Bits 38 to 35 represent opcode for operation 2. ▪ Bits 33 to 30 represent source1 for operation 2 ▪ Bits 28 to 25 represent source2 for operation 2 ▪ Bits 23 to 20 represent destination for operation 2 ▪ Bits 18 to 15 represent opcode for operation 3 ▪ Bits 13 to 10 represent source1 for operation 3 ▪ Bits 8 to 5 represent source2 for operation 3 ▪ Bits 3 to 0 represent destination for operation 3
data	input	192	This is a 192-bit data input to the VLIW microprocessor. Bits 191 to 128 represent data for operation 1, bits 127 to 64 represent data for operation 2, and bits 63 to 0 represents data for operation 3 of the VLIW instruction.
readdat-apipe1	output	64	Data output port for reading of data for operation 1 of VLIW instruction. When it is not reading data, the values are set to logic 0.
readdat-apipe2	output	64	Data output port for reading of data for operation 2 of VLIW instruction. When it is not reading data, the values are set to logic 0.
readdat-apipe3	output	64	Data output port for reading of data for operation 3 of VLIW instruction. When it is not reading data, the values are set to logic 0.
read-datavalid	output	1	This output signal is active high, indicating that the data at readdatapipe1 or readdatapipe2 or readdatapipe3 are valid.
jump	output	1	Output from VLIW microprocessor indicating that a branch has occurred and the instruction cache external to the VLIW microprocessor needs to fetch new instructions due to the branch.

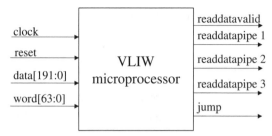

Figure 2.4   Diagram showing interface signals for VLIW microprocessor.

1. Instructions and data are fetched using an external instruction memory that has its own instruction cache. The defined VLIW microprocessor loads instructions and data directly from the external instruction memory through the 64-bit bus interface word and the 192-bit bus interface data. The output interface signal jump from the VLIW microprocessor is feedback as an input to the external instruction memory as an indicator that a branch has been taken and the instruction memory needs to pass another portion of instructions and data to the VLIW microprocessor.

2. The input signal clock to the VLIW microprocessor is generated from an external clock generator module.

3. The output bus readdatapipe1, readdatapipe2, readdatapipe3 is a 64-bit data bus that allows data to be read out of the VLIW microprocessor to external systems. The data are only valid when the output port readdatavalid is at logic 1.

4. The microprocessor does not have any register scoreboarding features within its shared register file.

Figure 2.5 shows the diagram of interfacing between the VLIW microprocessor with external systems.

2.3 Microarchitecture Specification

Section 2.2 describes the architecture for the VLIW microprocessor. The architecture shows the overall technical viewpoint of the design of the microprocessor. The next step after architectural specification is the microarchitectural definition. The architecture and microarchitecture are closely related as both are the starting points on which a design is defined.

In this step (microarchitecture specification), the block modules for the design are defined together with the top level intermodule signals. This step is viewed as the step in which a top level block diagram is defined. The following are considered for definition of the microarchitecture:

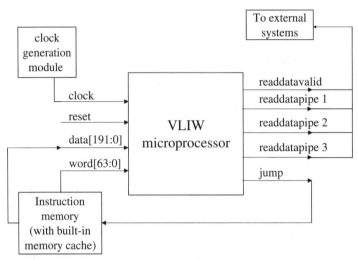

Figure 2.5 Diagram showing interface between VLIW microprocessor and external systems.

- Functional partitioning of the design
  - □ Blocks with similar functionality can be grouped together to form a design module. Designs with good functional partitioning allow the design to achieve optimal performance and gate utilization.
- Intermodule connectivity signals
  - □ Too many intermodule connectivity signals can complicate top level layout connection and may take up more area than necessary due to heavy congestion. However, this is dependent on the allowed area of layout and the fabrication process involved (the more layers the fabrication process allows, the better it is in handling congestion, but will increase the cost of fabrication).
- Intermodule signal naming
  - □ It is important to have a good naming convention in place in the design methodology. A proper naming convention allows the design to use proper names that are meaningful.
  - □ Having a proper naming convention is commonly overlookeded in a design project.
  - □ A good naming convention can be very useful during the design debug phase. It is ideal if a designer can obtain information on the design start-point, end-point, and active level (active high or low) of a signal just by its name.

Figure 2.6 shows the microarchitecture diagram for the VLIW microprocessor (figure drawn using Mentor Graphics' HDL Designer). The design is broken into five design modules:

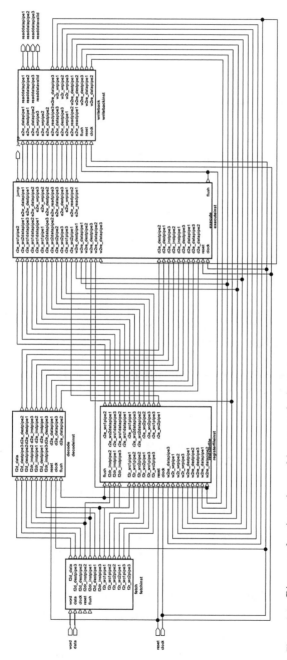

Figure 2.6 Diagram showing microarchitecture of microprocessor.

1. Module `fetch`—The functionality of this module is to fetch the necessary instructions and data from the external instruction memory module. The information is then passed to the `register file` module and `decode` module.

2. Module `decode`—In this module, the instructions fetched are decoded. It allows the VLIW microprocessor to "know" if the instruction is an `add`, `sub`, `mul`, `shift left`, `shift right`, or any of the other available operations. The information upon decoding is passed to the *execute* module.

3. Module `execute`—The decoded instruction is executed in this stage. It also receives data from the `register file` module to allow it to execute operations based on data from the internal registers. The result of the operation is passed to the `writeback` module.

4. Module `writeback`—In this module, the data computed by the `execute` module are written into the `register file` module for storage. Alternatively, the data can be output from this module to external systems for read operations.

5. Module `register file`—Register file module contains sixteen 64-bit registers which is used as internal storage for the VLIW microprocessor. When the `fetch` module has fetched an instruction from external instruction memory, it passes the information to the `register file`. This information allows the `register file` to pass the necessary data of its internal registers to the `execute` module. For example:

   ```
   add r0, r1, r2
   ```

   This operation requires the contents of register `r0` and `r1` to be added and stored into `r2`. The `register file` module passes the contents of `r0` and `r1` to the `execute` module. The results of the addition are passed to the `writeback` module and subsequently written into the `register file` module at `r2`.

Referring to the microarchitecture shown in Figure 2.10, the intermodule signals names are based on several simple rules:

- Each intermodule signal name is divided into two portions. Portion 1 and portion 2 of the name are separated by an underscore (_).
- Portion 1 of the signal name specifies where the signal came from and where the signal is heading.
- Portion 2 of the signal name represents the signal's true name.

- For example, the intermodule signal e2w_wrdatapipe1 is an output from module execute and input to module writeback (e2w represent the signal as an output from execute and an input to writeback). wrdatapipe1 is the name of the signal.

- Intermodule signal f2dr_instpipe1 is an output from fetch and input to decode and register file (f2dr represents the signal as an output from fetch and an input to decode and register file). instpipe1 is the name of the signal.

- The other signals that do not have portion 1 of the naming convention (signals that do not have f2d_, f2dr_, d2e_, e2w_, or w2r_) are the inputs and outputs of the VLIW microprocessor. For example, word, data, readdatapipe1, readdatapipe2, readdatapipe3, read-datavalid, jump, clock, and reset are input/output signals for the VLIW microprocessor.

Table 2.21 shows the description of the intermodule signals for the VLIW microprocessor.

TABLE 2.21 Description of Intermodule Signals for Microarchitecture of VLIW Microprocessor

Signal Name	Output From	Input To	Bits	Description
f2d_destpipe1	fetch	decode	4	Represents the destination register for operation 1 of a VLIW instruction
f2d_destpipe2	fetch	decode	4	Represents the destination register for operation 2 of a VLIW instruction
f2d_destpipe3	fetch	decode	4	Represents the destination register for operation 3 of a VLIW instruction
f2d_data	fetch	decode	192	192-bit data bus from the fetch module to the *decode* module
f2dr_instpipe1	fetch	decode, register file	4	Represents the instruction of operation 1
f2dr_instpipe2	fetch	decode, register file	4	Represents the instruction of operation 2
f2dr_instpipe3	fetch	decode, register file	4	Represents the instruction of operation 3
f2r_src1pipe1	fetch	register file	4	Represents the source1 register for operation 1

(*Continued*)

TABLE 2.21   Description of Intermodule Signals for Microarchitecture of VLIW Microprocessor (*continued*)

Signal Name	Output From	Input To	Bits	Description
f2r_src1pipe2	fetch	register file	4	Represents the source1 register for operation 2
f2r_src1pipe3	fetch	register file	4	Represents the source1 register for operation 3
f2r_src2pipe1	fetch	register file	4	Represents the source2 register for operation 1
f2r_src2pipe2	fetch	register file	4	Represents the source2 register for operation 2
f2r_src2pipe3	fetch	register file	4	Represents the source2 register for operation 3
d2e_instpipe1	decode	execute	4	Represents the instruction of operation 1
d2e_instpipe2	decode	execute	4	Represents the instruction of operation 2
d2e_instpipe3	decode	execute	4	Represents the instruction of operation 3
d2e_datapipe1	decode	execute	64	Represents the data for operation 1
d2e_datapipe2	decode	execute	64	Represents the data for operation 2
d2e_datapipe3	decode	execute	64	Represents the data for operation 3
d2e_destpipe1	decode	execute	4	Represents the *destination* register for operation 1
d2e_destpipe2	decode	execute	4	Represents the *destination* register for operation 2
d2e_destpipe3	decode	execute	4	Represents the *destination* register for operation 3
e2w_destpipe1	execute	writeback	4	Represents the *destination* register for operation 1
e2w_destpipe2	execute	writeback	4	Represents the *destination* register for operation 2
e2w_destpipe3	execute	writeback	4	Represents the *destination* register for operation 3
e2w_datapipe1	execute	writeback	64	Represents the computed data for operation 1 after the operation has executed
e2w_datapipe2	execute	writeback	64	Represents the computed data for operation 2 after the operation has executed
e2w_datapipe3	execute	writeback	64	Represents the computed data for operation 3 after the operation has executed

TABLE 2.21   Description of Intermodule Signals for Microarchitecture of VLIW Microprocessor (*continued*)

Signal Name	Output From	Input To	Bits	Description
e2w_wrpipe1	execute	writeback	1	Signifies to the writeback module that a write to register file is required for operation 1
e2w_wrpipe2	execute	writeback	1	Signifies to the writeback module that a write to register file is required for operation 2
e2w_wrpipe3	execute	writeback	1	Signifies to the writeback module that a write to register file is required for operation 3
e2w_readpipe1	execute	writeback	1	Signifies to the writeback module that a read to external system is required for operation 1
e2w_readpipe2	execute	writeback	1	Signifies to the writeback module that a read to external system is required for operation 2
e2w_readpipe3	execute	writeback	1	Signifies to the writeback module that a read to external system is required for operation 3
flush	execute	fetch, decode, writeback, register file	1	Global signal that flushes all the modules, indicating that a branch is to occur
w2r_wrpipe1	writeback	register file	1	This signal when valid represents writing of data from w2r_datapipe1 into register designated by w2r_destpipe1
w2r_wrpipe2	writeback	register file	1	This signal when valid represents writing of data from w2r_datapipe2 into register designated by w2r_destpipe2
w2r_wrpipe3	writeback	register file	1	This signal when valid represents writing of data from w2r_datapipe3 into register designated by w2r_destpipe3
w2re_destpipe1	writeback	register file, execute	4	Represents the destination register in the register file for a write on operation 1

(Continued)

TABLE 2.21   Description of Intermodule Signals for Microarchitecture of VLIW Microprocessor (*continued*)

Signal Name	Output From	Input To	Bits	Description
w2re_destpipe2	writeback	register file, execute	4	Represents the destination register in the register file for a write on operation 2
w2re_destpipe3	writeback	register file, execute	4	Represents the destination register in the register file for a write on operation 3
w2re_datapipe1	writeback	register file, execute	64	Represents the data to be written into the register designated by w2r_destpipe1
w2re_datapipe2	writeback	register file, execute	64	Represents the data to be written into the register designated by w2r_destpipe2
w2re_datapipe3	writeback	register file, execute	64	Represents the data to be written into the register designated by w2r_destpipe3
r2e_src1datapipe1	register file	execute	64	Represents the contents of register designated by f2r_src1pipe1; the data are passed to the execute module for execution of operation 1
r2e_src1datapipe2	register file	execute	64	Represents the contents of register designated by f2r_src1pipe2; the data are passed to the execute module for execution of operation 2
r2e_src1datapipe3	register file	execute	64	Represents the contents of register designated by f2r_src1pipe3; the data are passed to the execute module for execution of operation 3
r2e_src2datapipe1	register file	execute	64	Represent the contents of register designated by f2r_src2pipe1; the data are passed to the execute module for execution of operation 1
r2e_src2datapipe2	register file	execute	64	Represent the contents of register designated by f2r_src2pipe2; the data are passed to the execute module for execution of operation 2

**TABLE 2.21   Description of Intermodule Signals for Microarchitecture
of VLIW Microprocessor (*continued*)**

Signal Name	Output From	Input To	Bits	Description
r2e_src2datapipe3	register file	execute	64	Represent the contents of register designated by f2r_src2pipe3; the data are passed to the execute module for execution of operation 3
r2e_src1pipe1	register file	execute	4	Represents the source1 register in the register file for operation 1
r2e_src1pipe2	register file	execute	4	Represents the source1 register in the register file for operation 2
r2e_src1pipe3	register file	execute	4	Represents the source1 register in the register file for operation 3
r2e_src2pipe1	register file	execute	4	Represents the source2 register in the register file for operation 1
r2e_src2pipe2	register file	execute	4	Represents the source2 register in the register file for operation 2
r2e_src2pipe3	register file	execute	4	Represents the source2 register in the register file for operation 3

3

RTL Coding, Testbenching, and Simulation

Section 2.2 in Chapter 2 shows the architecture and Section 2.3 shows the microarchitecture of the VLIW microprocessor. Once the microarchitecture has been defined with the intermodule signals, the next step is to write the RTL code and testbenches to verify the code.

The RTL code is written based on the functionality of the design blocks or modules that are defined in the microarchitecture. For example, the fetch module will have the RTL code written for the functionality of fetching the VLIW instruction and data from external memory module to the decode module.

Note: RTL is register transfer level. RTL code refers to code that is written to reflect the functionality of a design. RTL code can be synthesized to logic gates using logic synthesis tools.

Note: There are three types of verilog code: structural, RTL, and behavioral. Structural verilog code describes the netlist of a design. An example of structural verilog is as follows:

AND   and_inst_0 (.O(abc), .I1(def), .I2(ghi));
OR   or_inst_3 (.O(xyz), .I1(kjl), .I2(mbp), .I3(hyf));

RTL code describes the functionality of a design and is synthesizable. Behavioral code describes the behavior of a design as a black box. It does not have details on how the functionality of a design is achieved, but rather a behavioral description of the design. Behavioral codes are normally used for verification and not for synthesis.

When writing the RTL code, it is important to follow a certain set of coding rules in order to have an efficient code that can synthesize to optimal solution. Different design centers normally have slightly different coding rules, but the objective of the coding rules is always the same: to achieve optimal synthesis results.

It is important to note that not all verilog syntax is synthesizable. Only a portion of verilog syntax can be synthesized. And synthesis results can vary greatly between a well written RTL code and an inefficient RTL code.

It is therefore important to have a good set of coding rules in place when writing verilog RTL code. Section 3.1 shows a set of coding rules that is used for the design of the VLIW microprocessor.

3.1   Coding Rules

The coding rules described in this chapter are a set of generic coding rules that can be used as a guideline to ensure good coding style as well as to obtain good verilog code to ensure optimal synthesis. Not having a good set of coding rules can result in badly coded RTL, which can cause a synthesis tool to synthesize redundant logic to a design. This will result in a greater number of logic gates. Alternately, the synthesis tool may also synthesize garbage logic, causing a mismatch between the RTL simulation and the synthesized logic circuit.

1. *Use comments in RTL code.* Many inexperienced designers often neglect putting comments into RTL code. This may cause difficulty when the RTL code is reused or reanalyzed at a later stage, because the original designer may have forgotten the reasons for the RTL code. Adding comments to a RTL code makes it readable and easier to understand. It is a good coding practice to always use comments when writing code.

2. *Module name matching filename.* Section 2.3 explained the advantages of using a naming convention for intermodule signals. Apart from the signals having a naming convention, it is good practice to ensure that the filename of the RTL code matches the module name of the code. Each filename should only have one RTL module. Following this rule makes the fullchip easily readable, especially when the fullchip is a large ASIC or SOC design that consists of many files.

3. *Output of each design module/block must be driven by a flip-flop (Figure 3.1).* Having a flip-flop at the output of each design module allows the timing path to end at the output, therefore simplifying the timing analysis of the design. Each flip-flop must also be resetable to ensure that during power up, the flip-flop can be reset to a known state.

The VLIW microprocessor consists of five design modules (fetch, decode, execute, writeback, and register file) as shown in Figure 2.6. Having a flip-flop at the output of each of the design

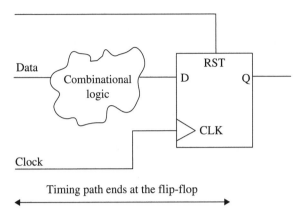

Timing path ends at the flip-flop

Figure 3.1   Diagram showing end of timing path at flip-flop.

module ensures that the timing path of each design modules ends at the output of each module. This simplifies the timing analysis of each module as the timing path is limited to only the corresponding module.

4. *Clock signal must be treated as a golden signal and no buffering is allowed in RTL code.* When writing RTL code to describe the functionality of a design, the clock input signal must be treated as golden, meaning the clock input signal cannot be coded to have any buffering. This is important because the control of clock skew in a design is done during clock tree synthesis (clock tree synthesis is part of auto-place-route). If any buffering is required, it is only allowed during clock tree synthesis (discussed in Section 3.13).

   Adding clock buffering into RTL code is inefficient and misleading because during RTL coding, the designer does not have accurate information on parasitic that is generated during layout. Therefore, adding clock buffers during the RTL coding stage is overkill as some of the buffering may not be necessary. Permitting clock buffering during clock tree synthesis allows much better control of clock skew.

5. *Gated clock should not be used unless necessary.* Gated clock is commonly used when designing for low power. Therefore, if a design is not meant for low power, clock gating should never be used in the RTL code. Use of gated clock in RTL code complicates the verification of the design because it may cause unnecessary glitches in the gated clock domain. Furthermore, gated clock complicates timing analysis. Gated clock is discussed in Section 3.13.

6. *It is important to define a reset as asynchronous or synchronous.* Asynchronous reset is a reset that can occur anytime while synchronous reset is a reset that can only occur during a valid clock. Table 3.1 shows the differences between asynchronous and synchronous resets.

TABLE 3.1   Differences between Asynchronous and Synchronous Reset

Asynchronous Reset	Synchronous Reset

``
// for active high reset
always @ (posedge clock or posedge reset)
begin
   if (reset)
      Q <= 0;
   else
      Q <= D;
end
``
	``
	// for active high reset
	always @ (posedge clock)
	begin
	   if (reset)
	      Q <= 0;
	   else
	      Q <= D;
	end
	``

Referring to the diagram below, output **Q** is reset to logic zero whenever reset is high, irrespective of clock.

Referring to the diagram below, output **Q** is reset to logic zero whenever reset is high during a rising edge of clock.

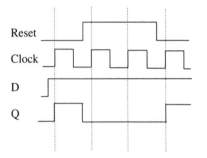

Asynchronous Active High Reset

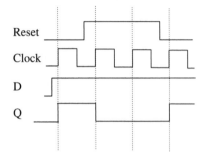

Synchronous Active High Reset

```
// for active low reset
always @ (posedge clock or negedge reset)
begin
   if (~reset)
      Q <= 0;
   else
      Q <= D;
end
```

```
//for active low reset
always @ (posedge clock)
begin
   if (~reset)
      Q <= 0;
   else
      Q <= D;
end
```

Referring to digram below, output **Q** is reset to logic zero whenever reset is low, irrespective of clock.

Referring to digram below, output **Q** is reset to logic zero whenever reset is low during a rising edge of clock

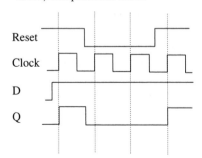

Asynchronous Active Low Reset

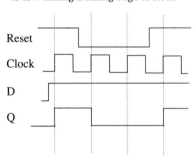

Synchronous Active Low Reset

7. *Usage of blocking and nonblocking statement.* Both of these statements are procedural statements used in `always` block. When an assignment occurs in blocking statement, the assignment is executed before proceeding to the next procedural statement. For a nonblocking statement, the assignment is scheduled at the end of the queue. The assignment execution occurs only at the end of the respective cycle. Table 3.2 shows an example of usage of blocking and nonblocking statements. The example shown for a nonblocking statement is correct while the use of a blocking statement is wrong.

Referring to the blocking statement of Table 3.2, changing the position of Q1 and Q2 in the procedural sequence will create different simulation results.

If the nonblocking statement in Table 3.2 is simulated, whether Q1 or Q2 is placed first in the procedural sequence, the simulation results are the same.

When writing RTL code, to ensure that simulation will never see any ambiguity in the simulation results, always follow these two rules when using nonblocking and blocking statements:

i. Use a nonblocking statement when writing code for sequential `always` block

```
always @ (posedge clock)
begin
        Q1 <= A & B;
end
```

ii. Use a blocking statement when writing code for combinational `always` block

```
always @ (A or B)
begin
        Q1 = A & B;
end
```

TABLE 3.2   **Wrong Use of Blocking Statement**

Nonblocking statement	Blocking statement

always @ (posedge clock)
  begin
    Q1 <= A & B;
    Q2 <= Q1 | C;
  end
	always @ (posedge clock)
	  begin
	    Q1 = A & B;
	    Q2 = Q1 | C;
	  end

> *Note*: A sequential "always block" refers to an always block that uses a `posedge clock` or `negedge clock` as its sensitivity list. When synthesized, a sequential always block will translate to rising edge flip-flop for `posedge clock` and a falling edge flip-flop for `negedge clock`.

> *Note*: A combinational "always block" is an always block that does not have negedge or posedge in its sensitivity list. It only uses signals in the always block list. When a combinational always block is synthesized, it will translate to combinational logic only.

For an indepth understanding on blocking and nonblocking statements, please refer to *Verilog Coding For Logic Synthesis* by Weng Fook Lee (John Wiley).

8. *Do not mix blocking and nonblocking statements in one always block.* Most synthesis tools do not allow an always block to have a combination of blocking and nonblocking statements. Although having such a combination allows the code to be simulated, synthesis will fail. It is a good coding practice to ensure that RTL code does not violate this rule!

9. *Avoid using* `initial` *statements in RTL code.* An `initial` statement is used to initialize values of signals in a verilog code. Some synthesis tools will fail when there are `initial` statements in the RTL code, while some synthesis tools will ignore the `initial` statements. However synthesis tools treat the initial statements, it should not be used in RTL coding. Usage of `initial` statements in RTL can cause problems during verification due to the mismatch between pre- and post-synthesis.

10. Using bitwise operator and logical operator. When bitwise operators are used on a bus, they operate on each bit of the bus and return the result in bus format. However, for logical operators the result of the operation is in TRUE or FALSE form. When a logical operator is used on a bus, the end result of the operation is a single bit TRUE or FALSE. Table 3.3 shows the difference between Logical operator and Bitwise operator.

11. *When using an* `if-else` *statement, ensure that unwanted latch is not inferred.* When using an `if-else` statement, all possible combinations must be specified. Alternatively, the `if-else` statement can use an *else* condition at the end of the `if-else` statement. This will ensure that unwanted latch is not inferred in the design. Table 3.4 shows the differences between a Complete and Incomplete `if-else` statement.

TABLE 3.3   Differences between Logical Operator and Bitwise Operator

Logical Operator	Bitwise Operator
| ```
module logical (A, B, C);
input [3:0] A, B;
output [3:0] C;
wire [3:0] C;
assign C = A && B;
endmodule
``` | ```
module bitwise (A, B, C);
input [3:0] A,B;
output [3:0] C;
wire [3:0] C;
assign C = A & B;
endmodule
``` |

A	1 0 1 1	A	1 0 1 1
B	0 1 1 1	B	0 1 1 1
C	0 0 0 1	C	0 0 1 1

Logical operator returns the value of the operation in TRUE or FALSE form. If the value of A is 1011, this translates to A being TRUE. If the value of B is 0111, this translates to B being TRUE. Therefore, the logical AND of A and B is TRUE:

C = A && B
  = TRUE && TRUE
  = TRUE

When synthesized, bits [3:1] of C will be tied to ground (logic zero). Only bit 0 is built using NOR gates. Synthesized circuit for module logical is shown in the diagram below.

Bitwise operator returns the value of the operation in bus form when the inputs are in bus form. If the value of A is 1011 and the value of B is 0111, the bitwise AND of A and B is 0011.

When synthesized, all 4 bits of output is built using an AND gate between bits of A and B. Synthesized circuit for module bitwise is shown in the diagram below.

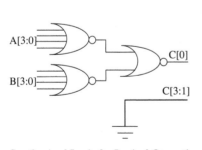

Synthesized Logic for Logical Operation

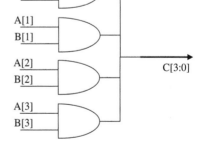

Synthesized Logic for Bitwise Operation

12. *When using a* case *statement, ensure that unwanted latch is not inferred.* When using a case statement, all possible combinations must be specified. Alternatively, the case statement can use a default condition at the end of the case statement. This will ensure that unwanted latch is not inferred in the design. Table 3.5 shows the differences between a Complete and Incomplete case statement.

TABLE 3.4 **Differences between a Complete and an Incomplete `if-else` Statement**

Complete `if-else` Statement	Incomplete `if-else` Statement

``
module complete (A,B,Q);
input A,B,C;
output Q;
reg Q;
always @ (A or B)
begin
  if (A & B)
    Q = 1;
  else if (~A & B)
    Q = 0;
  else if (A & ~B)
    Q = 0;
  else if (~A & ~B)
    Q = 0;
end
end module
``
	``
	module incomplete (A,B,Q);
	input A,B;
	output Q;
	reg Q;
	always @ (A or B)
	begin
	  if (A & B)
	    Q = 1;
	  else if (~A & B)
	    Q = 0;
	end
	endmodule
	``

All the different possible combination of A and B are specified in the if-else statement. The logic synthesized from the verilog code is an AND gate.

Only the combination of A=1, B=1 and A=0, B=1 are specified Since the other combination of A and B is not specified, during simulation the verilog simulator will maintain the previous value of Q if A=1, B=0 or A=0, B=0. As a result, if the incomplete verilog code is synthesized, the synthesis tool will infer a latch to enable output Q to hold its previous value. This is referred to as latch inference.

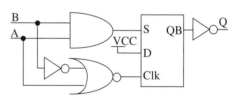

Synthesized Logic
Complete if-else
Statement

Synthesized Logic for Incomplete if-else
Statement

Referring to the diagram above, the logic synthesized is an AND gate. This is in accordance to the functionality of Q being generated from an AND function in verilog. Specifying all the possible combinations of A and B allows the synthesis tool to evaluate all the different combinations and synthesize the best logic that can fit the functionality. There may be occasions where it is difficult or tedious to specify all the possible combinations (especially for cases that involve many signals); use of keyword else at the end of the if-else statement is adequate.

Referring to the diagram above, a latch is inferred during synthesis as a means to allow output Q to maintain its previous value if A=1, B=0 or A=0, B=0. Therefore if a verilog code has an incomplete if-else statement, a latch is inferred. This is undesirable because it increases the amount of logic. To avoid latch inference, use a complete if-else statement that specifies all the different combination of A and B, or use the keyword else at the end of the if-else statement.

TABLE 3.4   Differences between a Complete and an Incomplete `if-else` Statement (Continued)

Complete if-else Statement	Incomplete if-else Statement

```
module useelse (A,B,Q);
input A,B;
output Q;
reg Q;
always @ (A or B)
begin
  if (A & B)
    Q = 1;
  else
    Q = 0;
end
endmodule
```

By using the `else` keyword, all other combinations of A and B that are not specified will assign the value of 0 to output Q. This ensures that a latch is not inferred even though not all possible combinations of A and B are specified. The logic synthesized for module `useelse` is the same as the AND logic shown in the diagram above.

13. *Partition a design such that each design module is between 5,000 gates and 50,000 gates.* A design can consist of many modules. Each module should be partitioned such that it should not be less than 5,000 gates or more than 50,000 gates. A design module that is partitioned too small would cause inefficient synthesis, while a design module that is partitioned too large will have very long synthesis run-time. Figure 3.2 shows a diagram of a design with poor partitioning with modules that are too small while other modules are too large.

    Figure 3.3 shows a design with good partitioning whereby all the design modules are within the range of 5,000 gates to 50,000 gates.

14. *Using X in coding for synthesis.* X is interpreted differently in simulation and in synthesis. During synthesis, an X represents a don't care, whereas during simulation, an X represents an unknown. When a synthesis tool encounters an X, the synthesis tool can perform optimization on the signal, as it is a don't care. Therefore, when coding for synthesis, RTL code can use X. However, use of X in RTL should be limited to internal signals and only used when necessary. Signals that are output of a design module should never be assigned with X's as these X's will propagate to other design modules during simulation and complicate simulation results analysis. Example 3.1 shows the RTL code that is used for coding a multiplexer with 2 bits `select`. In this example, output of the multiplexer is a don't care if `select` occurs at 11.

TABLE 3.5   Differences between a Complete and an Incomplete case Statement

Complete case Statement	Incomplete case Statement

module completecase (A,B,sel,Q);
input A,B;
input [1:0] sel;
output Q;
reg Q;
always @ (A or B or sel)
begin
   case (sel)
     2'b00: Q = A;
     2'b01: Q = B;
     2'b10: Q = 0;
     2'b11: Q = 0;
   endcase
end
endmodule
	module incompletecase (A,B,sel,Q);
	input A,B;
	input [1:0] sel;
	output Q;
	reg Q;
	always @ (A or B or sel)
	begin
	   case (sel)
	     2'b00: Q = A;
	     2'b01: Q = B;
	     2'b10: Q = 0;
	   endcase
	end
	endmodule

All the different combinations of sel are specified in the case statement. The logic synthesized from the verilog code uses combinational logic of a multiplexer and NOR gate to form the required functionality.	Only the combination of sel=00, sel=01, and sel=10 are specified. Because the remaining combination of sel=11 is not specified, during simulation the verilog simulator will maintain the previous value of Q if sel=11. If the verilog code is synthesized, the synthesis tool will infer a latch to enable output Q to hold its previous value for sel=11. This is referred to as latch inference.

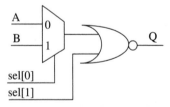

Synthesized Logic for Complete case Statement

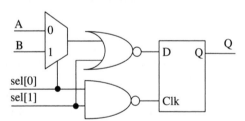

Synthesized Logic for Incomplete case Statement

Referring to the diagram above, the logic synthesized is a combination of a multiplexer and a NOR gate. Similarly to the if-else, there may be occasions where it is difficult or tedious to specify all the possible combinations (especially for cases that involve many signals); use of keyword default at the end of the case statement is adequate.

Referring to the diagram above, a latch is inferred during synthesis as a means to allow output Q to maintain its previous value if sel=11. This occurrence is similar to the latch inference for the incomplete if-else statement.

module defaultcase (A,B,sel,Q);
input A,B;
input [1:0] sel;
output Q;

**TABLE 3.5   Differences between a Complete and an Incomplete case Statement
(Continued)**

Complete case Statement	Incomplete case Statement

```
reg Q;
always @ (A or B or sel)
begin
   case (sel)
      2'b00:  Q = A;
      2'b01:  Q = B;
      default: Q = 0;
   endcase
end
endmodule
```

By using the default keyword, other combinations of sel that are not specified will assign the value of 0 to output Q. This ensures that a latch is not inferred even though not all possible combinations of sel are specified.

Example 3.1   Use of X for Synthesis

```
case (sel)
      2'b00: temp = A & B;
      2'b01: temp = B & C;
      2'b10: temp = C & D;
      default: temp = 1'bX;
endcase
```

15. *Avoid using infinite timing loop.* Infinite timing loops or "feedback" loops are combinational logic loops that have the output of the combinational logic being "feedback" to the input. Figure 3.4 shows an example of a combinational logic circuit having its output "feedback" to its input. This creates an infinite timing loop and complicates timing analysis. Designs with infinite timing loop on combinational logic must be avoided.

16. *Ensure that the sensitivity list is complete.* The sensitivity list is the list of signals that are used with an always block. Whenever a signal within the sensitivity list changes, the verilog code in the always block is evaluated by the simulator.

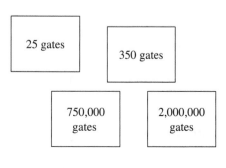

Figure 3.2   Diagram showing a design with poor partitioning.

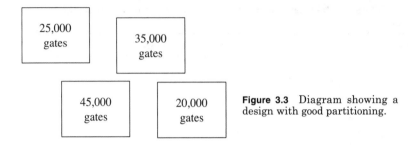

25,000 gates

35,000 gates

45,000 gates

20,000 gates

Figure 3.3 Diagram showing a design with good partitioning.

Example 3.2   Complete Sensitivity List for AND Function

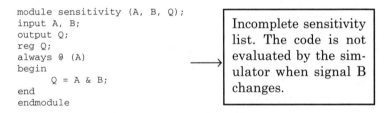

```
module sensitivity (A, B, Q);
input A, B;
output Q;
reg Q;
always @ (A or B)
begin
        Q = A & B;
end
endmodule
```

Complete sensitivity list. This `always` block is triggered whenever signal A or B changes.

Example 3.3 Shows the same verilog code as Example 3.2, but with an incomplete sensitivity list.

Example 3.3   Incomplete Sensitivity List for AND Function

```
module sensitivity (A, B, Q);
input A, B;
output Q;
reg Q;
always @ (A)
begin
        Q = A & B;
end
endmodule
```

Incomplete sensitivity list. The code is not evaluated by the simulator when signal B changes.

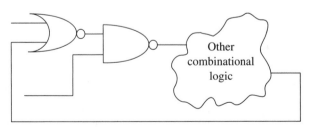

Other combinational logic

Figure 3.4   Diagram showing infinite timing loop of combinational logic.

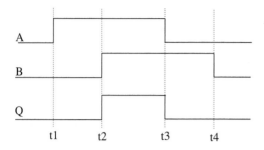

Figure 3.5 Simulation Waveform for verilog code with complete sensitivity list.

When the verilog codes of Example 3.2 and Example 3.3 are simulated, both will yield different simulation results due to the incomplete sensitivity list. Figure 3.5 shows the simulation waveform of Example 3.2.
Referring to Figure 3.5:

  a. At time t1, signal *A* changes. The sensitivity list for Example 3.2 is triggered and the evaluation of *Q* occurs. Since *B* is 0, *Q* is also 0.
  b. At time t2, signal *B* changes. The sensitivity list is triggered and the evaluation of *Q* occurs. Since *A* and *B* are 1, *Q* is 1.
  c. At time t3, signal *A* changes. The sensitivity list is triggered and the evaluation of *Q* occurs. Since *A* is 0, *Q* is 0.
  d. At time t4, signal *B* changes. The sensitivity list is triggered and the evaluation of *Q* occurs. Since *B* is 0, *Q* is 0.

Figure 3.6 shows the simulation waveform of Example 3.3.
Referring to Figure 3.6:

  a. At time t1, signal *A* changes. The sensitivity list for Example 3.3 is triggered and the evaluation of *Q* occurs. Since *B* is 0, *Q* is also 0.
  b. At time t2, signal *B* changes. However, signal *B* is not in the sensitivity list. Nothing occurs due to an incomplete sensitivity list.

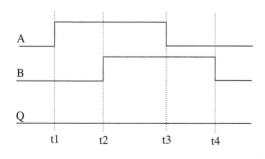

Figure 3.6 Simulation Waveform for verilog code with incomplete sensitivity list.

   c. At time t3, signal A changes. The sensitivity list is triggered and the evaluation of Q occurs. Since A is 0, Q is 0.

   d. At time t4, signal B changes. However, signal B is not in the sensitivity list. For Example 3.3, nothing occurs on Q due to an incomplete sensitivity list.

When simulating a verilog code, an incomplete sensitivity list will yield simulation results that do not accurately reflect the functionality of the RTL code.

However, when synthesized, both Example 3.2 and Example 3.3 will synthesize to an AND gate. But from simulation waveforms of Example 3.3, the output waveform does not reflect that of an AND gate, causing a mismatch between simulation and synthesis.

When writing RTL code, it is good practice to always ensure that the sensitivity list is complete to avoid mismatch.

3.2   RTL Coding

Section 3.1 shows a list of coding rules as a guideline when writing RTL code. These rules must be followed in order to obtain good RTL code that can translate to optimal synthesis results.

Referring to the architectural diagram of Figure 2.3 and microarchitectural diagram of Figure 2.6, the VLIW microprocessor consists of four stages (named `fetch`, `decode`, `execute` and `writeback`). For ease of understanding, each operation is numbered and categorized as pipe1, pipe2, and pipe3 with pipe1 operating operation 1, pipe2 operating operation 2, and pipe3 operating operation 3. All three operations within the VLIW instruction word have access to a sixteen 64-bit `register` file.

The RTL code for the VLIW microprocessor can be split into five separate modules: `fetch`, `decode`, `execute`, `writeback`, and `register` file (refer to Figure 2.6).

3.2.1   Module `fetch` RTL Code

The `fetch` module's functionality is to `fetch` VLIW instruction and data from an external instruction/data cache as shown in Figure 2.5. The fetched information is passed to the `decode` module to allow the instruction to be decoded. It is also passed to the `register` file module to allow the `execute` module to retrieve data from its register file for those operations that access internal registers.

Table 3.6 shows the interface signals for the `fetch` module and its interface signal functionality. Figure 3.7 shows the interface signal diagram of the `fetch` module.

TABLE 3.6   Interface Signals of *fetch* Module

Signal Name	Input/ Output	Bits	Description
word	Input	64	64-bit VLIW instruction word that represents three operations in parallel. Representation of each operation within the VLIW instruction word is shown in Section 2.1.1.
data	Input	192	The VLIW microprocessor processes three operations in parallel within one instruction word. The data provided for each operation are represented by this 192-bit data bus. As shown in Table 2.20, bits 191 to 128 represent data for operation 1, bits 127 to 64 represent data for operation 2, and bits 63 to 0 represent data for operation 3 of the VLIW instruction word.
clock	Input	1	Input clock pin. The VLIW microprocessor is active on rising edge of clock.
reset	Input	1	Input reset pin. Reset is asynchronous and active high.
flush	Input	1	This is a global signal that flushes all the modules, indicating that a branch is to occur.
f2d_data	Output	192	This is a 192-bit bus to pass the data fetched from external instruction memory to the decode unit.
f2d_destpipe1	Output	4	Represents the destination register for operation 1.
f2d_destpipe2	Output	4	Represents the destination register for operation 2.
f2d_destpipe3	Output	4	Represents the destination register for operation 3.
f2dr_instpipe1	Output	4	Represents the instruction of operation 1.
f2dr_instpipe2	Output	4	Represent the instruction of operation 2.
f2dr_instpipe3	Output	4	Represents the instruction of operation 3.
f2r_src1pipe1	Output	4	Represents the source1 register for operation 1.
f2r_src1pipe2	Output	4	Represents the source1 register for operation 2.
f2r_src1pipe3	Output	4	Represents the source1 register for operation 3.
f2r_src2pipe1	Output	4	Represents the source2 register for operation 1.
f2r_src2pipe2	Output	4	Represents the source2 register for operation 2.
f2r_src2pipe3	Output	4	Represents the source2 register for operation 3.

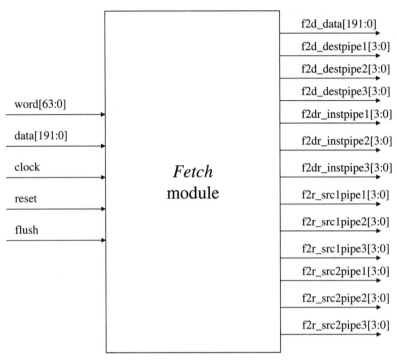

Figure 3.7 Diagram showing interface signals for fetch module.

Based on the interface signals shown in Table 3.6 with the signal functionality, the RTL verilog code for fetch module is shown in Example 3.4.

Example 3.4 RTL Verilog Code of fetch Module

```
module fetch (word, data, clock, reset, flush,
f2d_data,
f2d_destpipe1, f2d_destpipe2, f2d_destpipe3,
f2dr_instpipe1, f2dr_instpipe2, f2dr_instpipe3,
f2r_src1pipe1, f2r_src1pipe2, f2r_src1pipe3,
f2r_src2pipe1, f2r_src2pipe2, f2r_src2pipe3);

input clock; // clock input
input reset; // asynchronous reset active high
input flush; // when active high, the pipe is flushed
input [63:0] word; // this is for incoming VLIW
// instruction word
input [191:0] data; // incoming 192 bit data bus,
// 64 bits for each pipe

// represent the instruction for each pipe
output [3:0] f2dr_instpipe1, f2dr_instpipe2,
f2dr_instpipe3;
// represent the destination register for each
pipe
output [3:0] f2d_destpipe1, f2d_destpipe2, f2d_destpipe3;
```

→ Input port declaration

→ Output port declaration

```
// represent the source1 register for each pipe
output [3:0] f2r_src1pipe1, f2r_src1pipe2, f2r_src1pipe3;
// represent the source2 register for each pipe
output [3:0] f2r_src2pipe1, f2r_src2pipe2, f2r_src2pipe3;
// data bus output from fetch unit
output [191:0] f2d_data;

// include the file that declares the
// parameter declaration for
// register names and also instruction
// operations
    'include "regname.v"
```

```
                                        Refer to
                                        Section 3.2.1.2
```

```
reg [3:0] f2dr_instpipe1, f2dr_instpipe2, f2dr_instpipe3;
reg [3:0] f2r_src1pipe1, f2r_src1pipe2, f2r_src1pipe3;
reg [3:0] f2r_src2pipe1, f2r_src2pipe2, f2r_src2pipe3;
reg [3:0] f2d_destpipe1, f2d_destpipe2, f2d_destpipe3;
reg [191:0] f2d_data;

always @ (posedge clock or posedge reset)
begin
  // use non blocking for the following statements
  // within the posedge clock block
  if (reset)
  begin
    f2dr_instpipe1 <= nop;
    f2dr_instpipe2 <= nop;
    f2dr_instpipe3 <= nop;
    f2r_src1pipe1 <= reg0;
    f2r_src1pipe2 <= reg0;
    f2r_src1pipe3 <= reg0;
    f2r_src2pipe1 <= reg0;
    f2r_src2pipe2 <= reg0;
    f2r_src2pipe3 <= reg0;
    f2d_destpipe1 <= reg0;
    f2d_destpipe2 <= reg0;
    f2d_destpipe3 <= reg0;
    f2d_data <= 0;
  end
```

```
                                        Reset the
                                        signal to its
                                        default when
                                        reset occurs.
```

```
  else // positive edge clock detected
  begin
    if (~flush) // pipe is not being flushed
    begin
      // bits 64:60 are reserved
      // bits 59:40 are for operation 1
      // bits 39:20 are for operation 2
      // bits 19:0 are for operation 3
```

```
// fetch for operation 1
// bits 59, 54, 49, 44 are reserved
// bits 58:55 are for opcode
// bits 53:50 are for source1
// bits 48:45 are for source2
// bits 43:40 are for destination
case (word[58:55])
4'b0000:
  begin
    f2dr_instpipe1 <= nop;
  end
4'b0001:
```

```
                                        The 64-bit VLIW
                                        instruction word is fetched
                                        and passed on to decode
                                        module as instruction for
                                        operation 1.
```

```verilog
  begin
    f2dr_instpipe1 <= add;
  end
4'b0010:
  begin
    f2dr_instpipe1 <= sub;
  end
4'b0011:
  begin
    f2dr_instpipe1 <= mul;
  end
4'b0100:
  begin
    f2dr_instpipe1 <= load;
  end
4'b0101:
  begin
    f2dr_instpipe1 <= move;
  end
4'b0110:
  begin
    f2dr_instpipe1 <= read;
  end
4'b0111:
  begin
    f2dr_instpipe1 <= compare;
  end
4'b1000:
  begin
    f2dr_instpipe1 <= xorinst;
  end
4'b1001:
  begin
    f2dr_instpipe1 <= nandinst;
  end
4'b1010:
  begin
    f2dr_instpipe1 <= norinst;
  end
4'b1011:
  begin
    f2dr_instpipe1 <= notinst;
  end
4'b1100:
  begin
    f2dr_instpipe1 <= shiftleft;
  end
4'b1101:
  begin
    f2dr_instpipe1 <= shiftright;
  end
4'b1110:
  begin
    f2dr_instpipe1 <= bshiftleft;
  end
4'b1111:
  begin
    f2dr_instpipe1 <= bshiftright;
  end
default:
  begin
    f2dr_instpipe1 <= nop;
  end
endcase
```

```
case (word[53:50]) // for source1
register for pipe1
  4'b0000: f2r_src1pipe1 <= reg0;
  4'b0001: f2r_src1pipe1 <= reg1;
  4'b0010: f2r_src1pipe1 <= reg2;
  4'b0011: f2r_src1pipe1 <= reg3;
  4'b0100: f2r_src1pipe1 <= reg4;
  4'b0101: f2r_src1pipe1 <= reg5;
  4'b0110: f2r_src1pipe1 <= reg6;
  4'b0111: f2r_src1pipe1 <= reg7;
  4'b1000: f2r_src1pipe1 <= reg8;
  4'b1001: f2r_src1pipe1 <= reg9;
  4'b1010: f2r_src1pipe1 <= reg10;
  4'b1011: f2r_src1pipe1 <= reg11;
  4'b1100: f2r_src1pipe1 <= reg12;
  4'b1101: f2r_src1pipe1 <= reg13;
  4'b1110: f2r_src1pipe1 <= reg14;
  4'b1111: f2r_src1pipe1 <= reg15;
  default: f2r_src1pipe1 <= reg0;
endcase
```

The 64-bit VLIW instruction word is fetched and passed on to decode module as source1 register for operation 1.

```
case (word[48:45]) // for source2 register for pipe1
  4'b0000: f2r_src2pipe1 <= reg0;
  4'b0001: f2r_src2pipe1 <= reg1;
  4'b0010: f2r_src2pipe1 <= reg2;
  4'b0011: f2r_src2pipe1 <= reg3;
  4'b0100: f2r_src2pipe1 <= reg4;
  4'b0101: f2r_src2pipe1 <= reg5;
  4'b0110: f2r_src2pipe1 <= reg6;
  4'b0111: f2r_src2pipe1 <= reg7;
  4'b1000: f2r_src2pipe1 <= reg8;
  4'b1001: f2r_src2pipe1 <= reg9;
  4'b1010: f2r_src2pipe1 <= reg10;
  4'b1011: f2r_src2pipe1 <= reg11;
  4'b1100: f2r_src2pipe1 <= reg12;
  4'b1101: f2r_src2pipe1 <= reg13;
  4'b1110: f2r_src2pipe1 <= reg14;
  4'b1111: f2r_src2pipe1 <= reg15;
  default: f2r_src2pipe1 <= reg0;
endcase
```

The 64-bit VLIW instruction word is fetched and passed on to decode module as source2 register for operation 1.

```
case (word[43:40]) // for destination register // for pipe1
  4'b0000: f2d_destpipe1 <= reg0;
  4'b0001: f2d_destpipe1 <= reg1;
  4'b0010: f2d_destpipe1 <= reg2;
  4'b0011: f2d_destpipe1 <= reg3;
  4'b0100: f2d_destpipe1 <= reg4;
  4'b0101: f2d_destpipe1 <= reg5;
  4'b0110: f2d_destpipe1 <= reg6;
  4'b0111: f2d_destpipe1 <= reg7;
  4'b1000: f2d_destpipe1 <= reg8;
  4'b1001: f2d_destpipe1 <= reg9;
  4'b1010: f2d_destpipe1 <= reg10;
  4'b1011: f2d_destpipe1 <= reg11;
  4'b1100: f2d_destpipe1 <= reg12;
  4'b1101: f2d_destpipe1 <= reg13;
  4'b1110: f2d_destpipe1 <= reg14;
  4'b1111: f2d_destpipe1 <= reg15;
  default: f2d_destpipe1 <= reg0;
endcase
```

The 64-bit VLIW instruction word is fetched and passed on to decode module as destination register for operation 1.

```
// fetch for operation 2
// bits 39, 34, 29, 24
// bits 38:35 are for opcode
// bits 33:30 are for source1
// bits 28:25 are for source2
// bits 23:20 are for destination
case (word[38:35])
    4'b0000:
      begin
        f2dr_instpipe2 <= nop;
      end
    4'b0001:
      begin
        f2dr_instpipe2 <= add;
      end
    4'b0010:
      begin
        f2dr_instpipe2 <= sub;
      end
    4'b0011:
      begin
        f2dr_instpipe2 <= mul;
      end
    4'b0100:
      begin
        f2dr_instpipe2 <= load;
      end
    4'b0101:
      begin
        f2dr_instpipe2 <= move;
      end
    4'b0110:
      begin
        f2dr_instpipe2 <= read;
      end
    4'b0111:
      begin
        f2dr_instpipe2 <= compare;
      end
    4'b1000:
      begin
        f2dr_instpipe2 <= xorinst;
      end
    4'b1001:
      begin
        f2dr_instpipe2 <= nandinst;
      end
    4'b1010:
      begin
        f2dr_instpipe2 <= norinst;
      end
    4'b1011:
      begin
        f2dr_instpipe2 <= notinst;
      end
    4'b1100:
      begin
        f2dr_instpipe2 <= shiftleft;
      end
    4'b1101:
      begin
        f2dr_instpipe2 <= shiftright;
      end
```

The 64-bit VLIW instruction word is fetched and passed on to decode module as instruction for operation 2.

```
  4'b1110:
    begin
      f2dr_instpipe2 <= bshiftleft;
    end
  4'b1111:
    begin
      f2dr_instpipe2 <= bshiftright;
    end
  default:
    begin
      f2dr_instpipe2 <= nop;
    end
endcase
```

```
case (word[33:30]) // for source1
register for pipe2
  4'b0000: f2r_src1pipe2 <= reg0;
  4'b0001: f2r_src1pipe2 <= reg1;
  4'b0010: f2r_src1pipe2 <= reg2;
  4'b0011: f2r_src1pipe2 <= reg3;
  4'b0100: f2r_src1pipe2 <= reg4;
  4'b0101: f2r_src1pipe2 <= reg5;
  4'b0110: f2r_src1pipe2 <= reg6;
  4'b0111: f2r_src1pipe2 <= reg7;
  4'b1000: f2r_src1pipe2 <= reg8;
  4'b1001: f2r_src1pipe2 <= reg9;
  4'b1010: f2r_src1pipe2 <= reg10;
  4'b1011: f2r_src1pipe2 <= reg11;
  4'b1100: f2r_src1pipe2 <= reg12;
  4'b1101: f2r_src1pipe2 <= reg13;
  4'b1110: f2r_src1pipe2 <= reg14;
  4'b1111: f2r_src1pipe2 <= reg15;
  default: f2r_src1pipe2 <= reg0;
endcase
```

> The 64-bit VLIW instruction word is fetched and passed on to decode module as source1 register for operation 2.

```
case (word[28:25]) // for source2
register for pipe2
  4'b0000: f2r_src2pipe2 <= reg0;
  4'b0001: f2r_src2pipe2 <= reg1;
  4'b0010: f2r_src2pipe2 <= reg2;
  4'b0011: f2r_src2pipe2 <= reg3;
  4'b0100: f2r_src2pipe2 <= reg4;
  4'b0101: f2r_src2pipe2 <= reg5;
  4'b0110: f2r_src2pipe2 <= reg6;
  4'b0111: f2r_src2pipe2 <= reg7;
  4'b1000: f2r_src2pipe2 <= reg8;
  4'b1001: f2r_src2pipe2 <= reg9;
  4'b1010: f2r_src2pipe2 <= reg10;
  4'b1011: f2r_src2pipe2 <= reg11;
  4'b1100: f2r_src2pipe2 <= reg12;
  4'b1101: f2r_src2pipe2 <= reg13;
  4'b1110: f2r_src2pipe2 <= reg14;
  4'b1111: f2r_src2pipe2 <= reg15;
  default: f2r_src2pipe2 <= reg0;
endcase
```

> The 64-bit VLIW instruction word is fetched and passed on to decode module as source2 register for operation 2.

```
case (word[23:20]) // for destination register for // pipe2
  4'b0000: f2d_destpipe2 <= reg0;
  4'b0001: f2d_destpipe2 <= reg1;
  4'b0010: f2d_destpipe2 <= reg2;
  4'b0011: f2d_destpipe2 <= reg3;
  4'b0100: f2d_destpipe2 <= reg4;
```

```
    4'b0101: f2d_destpipe2 <= reg5;
    4'b0110: f2d_destpipe2 <= reg6;
    4'b0111: f2d_destpipe2 <= reg7;
    4'b1000: f2d_destpipe2 <= reg8;
    4'b1001: f2d_destpipe2 <= reg9;
    4'b1010: f2d_destpipe2 <= reg10;
    4'b1011: f2d_destpipe2 <= reg11;
    4'b1100: f2d_destpipe2 <= reg12;
    4'b1101: f2d_destpipe2 <= reg13;
    4'b1110: f2d_destpipe2 <= reg14;
    4'b1111: f2d_destpipe2 <= reg15;
    default: f2d_destpipe2 <= reg0;
  endcase
```

> The 64-bit VLIW instruction word is fetched and passed on to decode module as destination register for operation 2.

```
    // fetch for operation 3
    // bits 19, 14, 9, 4
    // bits 18:15 are for opcode
    // bits 13:10 are for source1
    // bits 8:5 are for source2
    // bits 3:0 are for destination
  case (word[18:15])
    4'b0000:
      begin
        f2dr_instpipe3 <= nop;
      end
    4'b0001:
      begin
        f2dr_instpipe3 <= add;
      end
    4'b0010:
      begin
        f2dr_instpipe3 <= sub;
      end
    4'b0011:
      begin
        f2dr_instpipe3 <= mul;
      end
    4'b0100:
      begin
        f2dr_instpipe3 <= load;
      end
    4'b0101:
      begin
        f2dr_instpipe3 <= move;
      end
    4'b0110:
      begin
        f2dr_instpipe3 <= read;
      end
    4'b0111:
      begin
        f2dr_instpipe3 <= compare;
      end
    4'b1000:
      begin
        f2dr_instpipe3 <= xorinst;
      end
    4'b1001:
      begin
        f2dr_instpipe3 <= nandinst;
      end
```

> The 64-bit VLIW instruction word is fetched and passed on to decode module as instruction for operation 3.

```
4'b1010:
  begin
    f2dr_instpipe3 <= norinst;
  end
4'b1011:
  begin
    f2dr_instpipe3 <= notinst;
  end
4'b1100:
  begin
    f2dr_instpipe3 <= shiftleft;
  end
4'b1101:
  begin
    f2dr_instpipe3 <= shiftright;
  end
4'b1110:
  begin
    f2dr_instpipe3 <= bshiftleft;
  end
4'b1111:
  begin
    f2dr_instpipe3 <= bshiftright;
  end
default:
  begin
    f2dr_instpipe3 <= nop;
  end
endcase

case (word[13:10]) // for source1
register for pipe3
  4'b0000: f2r_src1pipe3 <= reg0;
  4'b0001: f2r_src1pipe3 <= reg1;
  4'b0010: f2r_src1pipe3 <= reg2;
  4'b0011: f2r_src1pipe3 <= reg3;
  4'b0100: f2r_src1pipe3 <= reg4;
  4'b0101: f2r_src1pipe3 <= reg5;
  4'b0110: f2r_src1pipe3 <= reg6;
  4'b0111: f2r_src1pipe3 <= reg7;
  4'b1000: f2r_src1pipe3 <= reg8;
  4'b1001: f2r_src1pipe3 <= reg9;
  4'b1010: f2r_src1pipe3 <= reg10;
  4'b1011: f2r_src1pipe3 <= reg11;
  4'b1100: f2r_src1pipe3 <= reg12;
  4'b1101: f2r_src1pipe3 <= reg13;
  4'b1110: f2r_src1pipe3 <= reg14;
  4'b1111: f2r_src1pipe3 <= reg15;
  default: f2r_src1pipe3 <= reg0;
endcase

case (word[8:5]) // for source2
register for pipe3
  4'b0000: f2r_src2pipe3 <= reg0;
  4'b0001: f2r_src2pipe3 <= reg1;
  4'b0010: f2r_src2pipe3 <= reg2;
  4'b0011: f2r_src2pipe3 <= reg3;
  4'b0100: f2r_src2pipe3 <= reg4;
  4'b0101: f2r_src2pipe3 <= reg5;
  4'b0110: f2r_src2pipe3 <= reg6;
  4'b0111: f2r_src2pipe3 <= reg7;
  4'b1000: f2r_src2pipe3 <= reg8;
  4'b1001: f2r_src2pipe3 <= reg9;
```

The 64-bit VLIW instruction word is fetched and passed on to decode module as source1 register for operation 3.

The 64-bit VLIW instruction word is fetched and passed on to decode module as source2 register for operation 3.

```
  4'b1010: f2r_src2pipe3 <= reg10;
  4'b1011: f2r_src2pipe3 <= reg11;
  4'b1100: f2r_src2pipe3 <= reg12;
  4'b1101: f2r_src2pipe3 <= reg13;
  4'b1110: f2r_src2pipe3 <= reg14;
  4'b1111: f2r_src2pipe3 <= reg15;
  default: f2r_src2pipe3 <= reg0;
endcase

case (word[3:0]) // for destination register for // pipe3
  4'b0000: f2d_destpipe3 <= reg0;
  4'b0001: f2d_destpipe3 <= reg1;
  4'b0010: f2d_destpipe3 <= reg2;
  4'b0011: f2d_destpipe3 <= reg3;
  4'b0100: f2d_destpipe3 <= reg4;
  4'b0101: f2d_destpipe3 <= reg5;
  4'b0110: f2d_destpipe3 <= reg6;
  4'b0111: f2d_destpipe3 <= reg7;
  4'b1000: f2d_destpipe3 <= reg8;
  4'b1001: f2d_destpipe3 <= reg9;
  4'b1010: f2d_destpipe3 <= reg10;
  4'b1011: f2d_destpipe3 <= reg11;
  4'b1100: f2d_destpipe3 <= reg12;
  4'b1101: f2d_destpipe3 <= reg13;
  4'b1110: f2d_destpipe3 <= reg14;
  4'b1111: f2d_destpipe3 <= reg15;
  default: f2d_destpipe3 <= reg0;
endcase
```

> The 64-bit VLIW instruction word is fetched and passed on to decode module as destination register for operation 3.

```
if ((word[58:55] == 4'b0100) |
(word[38:35] ==
  4'b0100) | (word[18:15] == 4'b0100))
// load command
  f2d_data <= data;
else
  f2d_data <= 0;
end
```

> The 192-bit data bus is fetched and passed on to decode module if any of the operation is load.

```
else // flush
begin
    f2dr_instpipe1 <= nop;
    f2dr_instpipe2 <= nop;
    f2dr_instpipe3 <= nop;
    f2r_src1pipe1 <= reg0;
    f2r_src1pipe2 <= reg0;
    f2r_src1pipe3 <= reg0;
    f2r_src2pipe1 <= reg0;
    f2r_src2pipe2 <= reg0;
    f2r_src2pipe3 <= reg0;
    f2d_destpipe1 <= reg0;
    f2d_destpipe2 <= reg0;
    f2d_destpipe3 <= reg0;
    f2d_data <= 0;
  end
  end
end
endmodule
```

> Reset the signals to its default when flushing occurs.

The fetch module's functionality is to fetch the VLIW instruction and pass it to following stage. However, referring to Example 3.4 of the RTL verilog code, some of the logic in fetch module is used for decoding the fetched VLIW instruction. An example is the decoding of word[63:0] to form source1, source2, destination, and instruction. This decoding logic is put in the fetch module and not in the decode module to enable sharing of decoding logic between the fetch module and the decode module. Commonly all decoding logic is located in the decode module. However, putting too much logic in one module will slow the module's performance, thereby hindering the performance of the VLIW microprocessor. Therefore, some of the decoding logic is brought forward to the fetch module, allowing both the fetch and decode module's to share the responsibility of decoding. This in turn balances the critical path between the fetch module and decode module, allowing better overall speed performance. This concept is referred as register/logic balancing.

3.2.1.1 Register/logic Balancing
Register/logic balancing is a method used in design to balance the amount of logic between several register stages to achieve optimal performance of a design. Most synthesis tools like Synopsys's Design Compiler have built-in synthesis commands that allow a synthesis tool to perform logic balancing between different levels of register stages.

Figure 3.8 shows a logic circuit that cannot meet clock specification. The clock is targeted to run at 100 MHz or 10 ns per clock period. However, the first set of logic in the circuit has a total propagation delay of 13ns while the second set of logic has a total propagation delay of 2 ns. This creates a negative slack of 3 ns.

> *Note*: Negative slack occurs when a design is not able to meet timing specification. A design with negative slack of 3 ns is a design that cannot meet timing specification by 3 ns.

To optimize the design shown in Figure 3.8, the concept of register/logic balancing is used to balance the logic between both stages

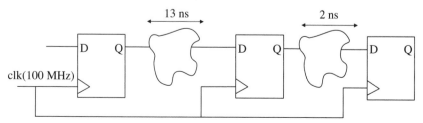

Figure 3.8 Diagram showing a critical path design.

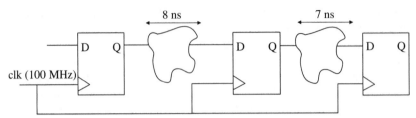

Figure 3.9 Diagram showing the critical path design with register balancing.

of the circuit. Because the first set of logic requires 13 ns while the second stage of logic requires only 2 ns, some of the logic from the first set can be moved to the second set. This allows the logic functionality of the design to be shared equally among both sets of logic, allowing for an optimized design.

Figure 3.9 shows the design after register balancing to share the functionality between both sets of logic gates. By sharing the logic, the first set of logic gates has a propagation delay of 8 ns while the second set of logic gates has a propagation delay of 7 ns, allowing the design to achieve the specified 100-MHz clock speed.

3.2.1.2 Contents of `regname.v` Verilog File The RTL verilog code of Example 3.4 includes a separate verilog file `regname.v`. The contents of `regname.v` declare two lists of names. One is for representation of the register names as `reg0`, `reg1`, `reg2`, ... `reg15`, while the second represents the operations as `nop`, `add`, `sub`, ... `bshiftright`, instead of hexadecimal numbers. This allows for simple representation using names rather than numbers. The verilog code of `regname.v` is shown in Example 3.5.

Example 3.5 Verilog Code for `regname.v`

```
parameter [3:0] reg0 = 4'h0,
    reg1 = 4'h1,
    reg2 = 4'h2,
    reg3 = 4'h3,
    reg4 = 4'h4,
    reg5 = 4'h5,
    reg6 = 4'h6,
    reg7 = 4'h7,
    reg8 = 4'h8,
    reg9 = 4'h9,
    reg10 = 4'ha,
    reg11 = 4'hb,
    reg12 = 4'hc,
    reg13 = 4'hd,
    reg14 = 4'he,
    reg15 = 4'hf;
parameter [3:0] nop = 4'h0,
    add = 4'h1,
    sub = 4'h2,
```

```
mul = 4'h3,
load = 4'h4,
move = 4'h5,
read = 4'h6,
compare = 4'h7,
xorinst = 4'h8,
nandinst = 4'h9,
norinst = 4'ha,
notinst = 4'hb,
shiftleft = 4'hc,
shiftright = 4'hd,
bshiftleft = 4'he,
bshiftright = 4'hf;
```

3.2.2 Module decode RTL Code

The decode module's functionality is to decode the operation passed from the fetch module. The operation is passed to execute module for execution. Table 3.7 shows the interface signals for the decode module and its functionality. Figure 3.10 shows the interface signal diagram of the decode module.

The RTL verilog code for decode module is shown in Example 3.6.

Example 3.6 RTL Verilog Code of decode Module

```
module decode (
f2d_destpipe1, f2d_destpipe2, f2d_destpipe3,
f2dr_instpipe1, f2dr_instpipe2, f2dr_instpipe3,
d2e_instpipe1, d2e_instpipe2, d2e_instpipe3,
d2e_destpipe1, d2e_destpipe2, d2e_destpipe3,
d2e_datapipe1, d2e_datapipe2, d2e_datapipe3,
clock, reset, flush, f2d_data);
```

```
input [3:0] f2d_destpipe1,
f2d_destpipe2, f2d_destpipe3;          ───→  Input port declaration
input [3:0] f2dr_instpipe1,
f2dr_instpipe2, f2dr_instpipe3;

input [191:0] f2d_data; // 192 bits
// data, 64 bit each pipe
input clock, flush, reset;
```

```
output [3:0] d2e_destpipe1,
d2e_destpipe2, d2e_destpipe3;          ───→  Output port declaration
output [3:0] d2e_instpipe1,
d2e_instpipe2, d2e_instpipe3;
output [63:0] d2e_datapipe1, d2e_datapipe2, d2e_datapipe3;
```

```
reg [3:0] d2e_destpipe1, d2e_destpipe2, d2e_destpipe3;
reg [3:0] d2e_instpipe1, d2e_instpipe2, d2e_instpipe3;
reg [63:0] d2e_datapipe1, d2e_datapipe2, d2e_datapipe3;
```

```
// include the file that declares the
// parameter declaration for
// register names and also            ───→  Refer to Section 3.2.1.2
// instruction operations
`include "regname.v"
```

TABLE 3.7   Interface Signals of decode Module

Signal Name	Input/ Output	Bits	Description
clock	Input	1	Input clock pin. The VLIW microprocessor is active on rising edge of clock.
reset	Input	1	Input reset pin. Reset is asynchronous and active high.
flush	Input	1	This is a global signal that flushes all the modules, indicating that a branch is to occur.
f2d_data	Input	192	This is a 192-bit bus to pass the data fetched from external instruction memory to the decode unit.
f2d_destpipe1	Input	4	Represents the destination register for operation 1.
f2d_destpipe2	Input	4	Represents the destination register for operation 2.
f2d_destpipe3	Input	4	Represents the destination register for operation 3.
f2dr_instpipe1	Input	4	Represents the instruction of operation 1.
f2dr_instpipe2	Input	4	Represents the instruction of operation 2.
f2dr_instpipe3	Input	4	Represents the instruction of operation 3.
d2e_instpipe1	Output	4	Represents the instruction of operation 1.
d2e_instpipe2	Output	4	Represents the instruction of operation 2.
d2e_instpipe3	Output	4	Represents the instruction of operation 3.
d2e_destpipe1	Output	4	Represents the destination register for operation 1.
d2e_destpipe2	Output	4	Represents the destination register for operation 2.
d2e_destpipe3	Output	4	Represents the destination register for operation 3.
d2e_datapipe1	Output	64	Represents the data for operation 1 of the VLIW instruction. The data bus is used only during load instruction.
d2e_datapipe2	Output	64	Represents the data for operation 2 of the VLIW instruction. The data bus is used only during load instruction.
d2e_datapipe3	Output	64	Represents the data for operation 3 of the VLIW instruction. The data bus is used only during load instruction.

```
always @ (posedge clock or posedge reset)
begin
  if (reset)
  begin
    d2e_instpipe1 <= nop;
    d2e_instpipe2 <= nop;
    d2e_instpipe3 <= nop;
    d2e_destpipe1 <= reg0;
    d2e_destpipe2 <= reg0;
    d2e_destpipe3 <= reg0;
    d2e_datapipe1 <= 0;

    d2e_datapipe2 <= 0;
    d2e_datapipe3 <= 0;
  end
```

⟶ Reset the signals to its default when reset occurs.

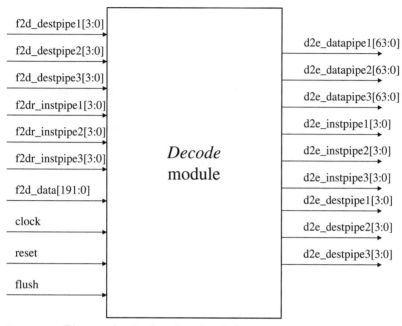

Figure 3.10 Diagram showing interface signals for decode module.

```
else // positive edge of clock detected
begin
if (~flush)
begin
    // decode for operation 1, pipe1
    case (f2dr_instpipe1)
      nop:
        begin
        // in no operation, all default to
        // zero
        d2e_destpipe1 <= reg0;
        d2e_datapipe1 <= 0;
      end
    add:
      begin
      // add src1, src2, dest -> src1 +
      // src2
      d2e_destpipe1 <= f2d_destpipe1;
      d2e_datapipe1 <= 0;
      end
    sub:
      begin
      // sub src1, src2, dest -> src1 -
      // src2
      d2e_destpipe1 <= f2d_destpipe1;
      d2e_datapipe1 <= 0;
      end
    mul:
      begin
      // mul src1, src2, dest
      // -> src1 x src2.
      // only 32 bits of src1, src2
```

Passing of desti-
nation register

```
      // considered
      d2e_destpipe1 <= f2d_destpipe1;
      d2e_datapipe1 <= 0;
   end
load:
   begin
      // load data from data bus to dest
      d2e_destpipe1 <= f2d_destpipe1;
      d2e_datapipe1 <= f2d_data[191:128];
      // operation 1 data load
   end
move:
   begin
      // move contents from src1 to dest
      d2e_destpipe1 <= f2d_destpipe1;
      d2e_datapipe1 <= 0;
   end
read:
   begin
      // read data to dest
      d2e_destpipe1 <= f2d_destpipe1;
      d2e_datapipe1 <= 0;
   end
compare:
   begin
      // compare src1, src2, dest
      // results of comparison stored in
      // dest
      d2e_destpipe1 <= f2d_destpipe1;
      d2e_datapipe1 <= 0;
   end
xorinst:
   begin
      // xorinst src1, src2, dest
      d2e_destpipe1 <= f2d_destpipe1;
      d2e_datapipe1 <= 0;
   end
nandinst:
   begin
      // nandinst src1, src2, dest
      d2e_destpipe1 <= f2d_destpipe1;
      d2e_datapipe1 <= 0;
   end
norinst:
   begin
      // norinst src1, src2, dest
      d2e_destpipe1 <= f2d_destpipe1;
      d2e_datapipe1 <= 0;
   end
notinst:
   begin
      // notinst src1, dest
      d2e_destpipe1 <= f2d_destpipe1;
      d2e_datapipe1 <= 0;
   end
shiftleft:
   begin
      // shiftleft src1, src2, dest
      d2e_destpipe1 <= f2d_destpipe1;
      d2e_datapipe1 <= 0;
   end
```

> Passing of data for load instruction of operation 1

```
shiftright:
  begin
    // shiftright src1, src2, dest
    d2e_destpipe1 <= f2d_destpipe1;
    d2e_datapipe1 <= 0;
  end
bshiftleft:
  begin
    // bshiftleft left src1, src2, dest
    d2e_destpipe1 <= f2d_destpipe1;
    d2e_datapipe1 <= 0;
  end
bshiftright:
  begin
    // bshiftright src1, src2, dest
    d2e_destpipe1 <= f2d_destpipe1;
    d2e_datapipe1 <= 0;
  end
default:
  begin
    // default
    d2e_destpipe1 <= reg0;
    d2e_datapipe1 <= 0;
  end
endcase

// decode for operation 2, pipe2
case (f2dr_instpipe2)
  nop:
    begin
      // in no operation, all default to
      // zero
      d2e_destpipe2 <= reg0;
      d2e_datapipe2 <= 0;
    end
add:
  begin
    // add src1, src2, dest -> src1 +
    // src2
    d2e_destpipe2 <= f2d_destpipe2;
    d2e_datapipe2 <= 0;
  end
sub:
  begin
    // sub src1, src2, dest -> src1 -
    // src2
    d2e_destpipe2 <= f2d_destpipe2;
    d2e_datapipe2 <= 0;
  end
mul:
  begin
    // mul src1, src2, dest
    // -> src1 x src2
    // only 32 bits of src1, src2
    // considered
    d2e_destpipe2 <= f2d_destpipe2;
    d2e_datapipe2 <= 0;
  end
load:
  begin
    // load data from data bus to dest
    d2e_destpipe2 <= f2d_destpipe2;
    d2e_datapipe2 <= f2d_data[127:64];
```

→ Passing of destination register

→ Passing of data for load instruction of operation 2

```
      // operation 2 data load
   end
move:
  begin
    // move contents from src1 to dest
    d2e_destpipe2 <= f2d_destpipe2;
    d2e_datapipe2 <= 0;
  end
read:
  begin
    // read data to dest
    d2e_destpipe2 <= f2d_destpipe2;
    d2e_datapipe2 <= 0;
  end
compare:
  begin
    // compare src1, src2, dest
    // results of comparison stored in
    // destination
    d2e_destpipe2 <= f2d_destpipe2;
    d2e_datapipe2 <= 0;
  end
xorinst:
  begin
    // xorinst src1, src2, dest
    d2e_destpipe2 <= f2d_destpipe2;
    d2e_datapipe2 <= 0;
  end
nandinst:
  begin
    // nandinst src1, src2, dest
    d2e_destpipe2 <= f2d_destpipe2;
    d2e_datapipe2 <= 0;
  end
norinst:
  begin
    // norinst src1, src2, dest
    d2e_destpipe2 <= f2d_destpipe2;
    d2e_datapipe2 <= 0;
  end
notinst:
  begin
    // notinst src1, dest
    d2e_destpipe2 <= f2d_destpipe2;
    d2e_datapipe2 <= 0;
  end
shiftleft:
  begin
    // shiftleft src1, src2, dest
    d2e_destpipe2 <= f2d_destpipe2;
    d2e_datapipe2 <= 0;
  end
shiftright:
  begin
    // shiftright src1, src2, dest
    d2e_destpipe2 <= f2d_destpipe2;
    d2e_datapipe2 <= 0;
  end
bshiftleft:
  begin
    // bshiftleft left src1, src2, dest
    d2e_destpipe2 <= f2d_destpipe2;
```

```
      d2e_datapipe2 <= 0;
    end
bshiftright:
  begin
    // bshiftright src1, src2, dest
    d2e_destpipe2 <= f2d_destpipe2;
    d2e_datapipe2 <= 0;
  end
default:
  begin
    // default
    d2e_destpipe2 <= reg0;
    d2e_datapipe2 <= 0;
  end
    endcase
  // decode for operation 3, pipe3
  case (f2dr_instpipe1)
nop:
  begin
    // in no operation, all default to
    // zero
    d2e_destpipe3 <= reg0;
    d2e_datapipe3 <= 0;
  end
add:
  begin
    // add src1, src2, dest -> src1 +
    // src2
    d2e_destpipe3 <= f2d_destpipe3;
    d2e_datapipe3 <= 0;
  end
sub:
  begin
    // sub src1, src2, dest -> src1 -
    // src2
    d2e_destpipe3 <= f2d_destpipe3;
    d2e_datapipe3 <= 0;
  end
mul:
  begin
    // mul src1, src2, dest
    // -> src1 x src2
    // only 32 bits of src1, src2
    // considered
    d2e_destpipe3 <= f2d_destpipe3;
    d2e_datapipe3 <= 0;
  end
load:
  begin
    // load data from data bus to
    dest
    d2e_destpipe3 <= f2d_destpipe3;
    d2e_datapipe3 <= f2d_data[63:0];
    // operation 3 data load
  end
move:
  begin
    // move contents from src1 to dest
    d2e_destpipe3 <= f2d_destpipe3;
    d2e_datapipe3 <= 0;
  end
```

Passing of destination register

Passing of data for load instruction of operation 3

```
read:
 begin
   // read data to dest
   d2e_destpipe3 <= f2d_destpipe3;
   d2e_datapipe3 <= 0;
 end
compare:
 begin
   // compare src1, src2, dest
   // results of comparison stored in
   // destination
   d2e_destpipe3 <= f2d_destpipe3;
   d2e_datapipe3 <= 0;
 end
xorinst:
 begin
   // xorinst src1, src2, dest
   d2e_destpipe3 <= f2d_destpipe3;
   d2e_datapipe3 <= 0;
 end
nandinst:
 begin
   // nandinst src1, src2, dest
   d2e_destpipe3 <= f2d_destpipe3;
   d2e_datapipe3 <= 0;
 end
norinst:
 begin
   // norinst src1, src2, dest
   d2e_destpipe3 <= f2d_destpipe3;
   d2e_datapipe3 <= 0;
 end
notinst:
 begin
   // notinst src1, dest
   d2e_destpipe3 <= f2d_destpipe3;
   d2e_datapipe3 <= 0;
 end
shiftleft:
 begin
   // shiftleft src1, src2, dest
   d2e_destpipe3 <= f2d_destpipe3;
   d2e_datapipe3 <= 0;
 end
shiftright:
 begin
   // shiftright src1, src2, dest
   d2e_destpipe3 <= f2d_destpipe3;
   d2e_datapipe3 <= 0;
 end
bshiftleft:
 begin
   // bshiftleft left src1, src2, dest
   d2e_destpipe3 <= f2d_destpipe3;
   d2e_datapipe3 <= 0;
 end
bshiftright:
 begin
   // bshiftright src1, src2, dest
   d2e_destpipe3 <= f2d_destpipe3;
   d2e_datapipe3 <= 0;
 end
```

```
        default:
          begin
            // default
            d2e_destpipe3 <= reg0;
            d2e_datapipe3 <= 0;
          end
      endcase
      d2e_instpipe1 <= f2dr_instpipe1;
      d2e_instpipe2 <= f2dr_instpipe2;
      d2e_instpipe3 <= f2dr_instpipe3;
    end
    else // flush
    begin
        // flushing causing all set to default
        d2e_instpipe1 <= nop;
        d2e_instpipe2 <= nop;
        d2e_instpipe3 <= nop;
        d2e_destpipe1 <= reg0;
        d2e_destpipe2 <= reg0;
        d2e_destpipe3 <= reg0;
        d2e_datapipe1 <= 0;
        d2e_datapipe2 <= 0;
        d2e_datapipe3 <= 0;
      end
    end
  end
endmodule
```

The output signals d2e_datapipe1, d2e_datapipe2, d2e_datapipe3 are always set to zero except for the load instruction. The output signals d2e_destpipe1, d2e_destpipe2, d2e_destpipe3 are always the flopped version of input signals f2d_destpipe1, f2d_destpipe2, f2d_destpipe3 for all instruction except for nop instruction. This allows for simplifying the verilog RTL code of Example 3.6 to that shown in Example 3.7. Both Example 3.6 and Example 3.7 are the same functionally and synthesize to the same logic.

Example 3.7 Simplified Verilog Code of Example 3.6

```
module decode (
f2d_destpipe1, f2d_destpipe2, f2d_destpipe3,
f2dr_instpipe1, f2dr_instpipe2, f2dr_instpipe3,
d2e_instpipe1, d2e_instpipe2, d2e_instpipe3,
d2e_destpipe1, d2e_destpipe2, d2e_destpipe3,
d2e_datapipe1, d2e_datapipe2, d2e_datapipe3,
clock, reset, flush, f2d_data
);

input [3:0] f2d_destpipe1, f2d_destpipe2, f2d_destpipe3;
input [3:0] f2dr_instpipe1, f2dr_instpipe2, f2dr_instpipe3;
input [191:0] f2d_data; // 192 bits data, 64 bit each pipe
input clock, flush, reset;

output [3:0] d2e_destpipe1, d2e_destpipe2, d2e_destpipe3;
output [3:0] d2e_instpipe1, d2e_instpipe2, d2e_instpipe3;
output [63:0] d2e_datapipe1, d2e_datapipe2, d2e_datapipe3;

reg [3:0] d2e_destpipe1, d2e_destpipe2, d2e_destpipe3;
reg [3:0] d2e_instpipe1, d2e_instpipe2, d2e_instpipe3;
```

```verilog
reg [63:0] d2e_datapipe1, d2e_datapipe2, d2e_datapipe3;
// include the file that declares the parameter declaration for register
// names and also instruction operations
'include "regname.v"

always @ (posedge clock or posedge reset)
begin
  if (reset)
  begin
    d2e_instpipe1 <= nop;
    d2e_instpipe2 <= nop;
    d2e_instpipe3 <= nop;
    d2e_destpipe1 <= reg0;
    d2e_destpipe2 <= reg0;
    d2e_destpipe3 <= reg0;
    d2e_datapipe1 <= 0;
    d2e_datapipe2 <= 0;
    d2e_datapipe3 <= 0;
  end
  else // positive edge of clock detected
  begin
    if (~flush)
    begin
      if (f2dr_instpipe1 == load)
        d2e_datapipe1 <= f2d_data[191:128];
      else
        d2e_datapipe1 <= 0;

      if (f2dr_instpipe2 == load)
        d2e_datapipe2 <= f2d_data[127:64];
      else
        d2e_datapipe2 <= 0;

      if (f2dr_instpipe3 == load)
        d2e_datapipe3 <= f2d_data[63:0];
      else
        d2e_datapipe3 <= 0;

      if (f2dr_instpipe1 == nop)
        d2e_destpipe1 <= reg0;
      else
        d2e_destpipe1 <= f2d_destpipe1;

      if (f2dr_instpipe2 == nop)
        d2e_destpipe2 <= reg0;
      else
        d2e_destpipe2 <= f2d_destpipe2;

      if (f2dr_instpipe3 == nop)
        d2e_destpipe3 <= reg0;
      else
        d2e_destpipe3 <= f2d_destpipe3;
      d2e_instpipe1 <= f2dr_instpipe1;
      d2e_instpipe2 <= f2dr_instpipe2;
      d2e_instpipe3 <= f2dr_instpipe3;
    end
    else // flush
    begin
      // flushing causing all set to default
      d2e_instpipe1 <= nop;
      d2e_instpipe2 <= nop;
      d2e_instpipe3 <= nop;
```

```
        d2e_destpipe1 <= reg0;
        d2e_destpipe2 <= reg0;
        d2e_destpipe3 <= reg0;
        d2e_datapipe1 <= 0;
        d2e_datapipe2 <= 0;
        d2e_datapipe3 <= 0;
    end
  end
end
endmodule
```

3.2.3 Module *register file* RTL Code

The register file module's functionality is to act as a local storage space in the VLIW microprocessor. Contents of the register file module is read and passed to the execute module, while results of operations is written to the register file module by the writeback module.

To maintain simplicity and ease of understanding on the register file module, the function of register scoreboarding is not implemented.

Table 3.8 shows the interface signals for the register file module and its interface signal functionality. Figure 3.11 shows the interface signal diagram of the register file module.

Based on the interface signals shown in Table 3.8 with the signal functionality, the RTL verilog code for register file module is shown in Example 3.8.

Example 3.8 RTL Verilog Code of register file Module

```
module registerfile (
clock, flush, reset,
f2r_src1pipe1, f2r_src1pipe2, f2r_src1pipe3,
f2r_src2pipe1, f2r_src2pipe2, f2r_src2pipe3,
f2dr_instpipe1, f2dr_instpipe2, f2dr_instpipe3,
w2re_datapipe1, w2re_datapipe2, w2re_datapipe3,
w2r_wrpipe1, w2r_wrpipe2, w2r_wrpipe3,
w2re_destpipe1, w2re_destpipe2, w2re_destpipe3,
r2e_src1datapipe1, r2e_src1datapipe2, r2e_src1datapipe3,
r2e_src2datapipe1, r2e_src2datapipe2, r2e_src2datapipe3,
r2e_src1pipe1, r2e_src1pipe2, r2e_src1pipe3,
r2e_src2pipe1, r2e_src2pipe2, r2e_src2pipe3
);

input [3:0] f2r_src1pipe1, f2r_src1pipe2, f2r_src1pipe3;
input [3:0] f2r_src2pipe1, f2r_src2pipe2, f2r_src2pipe3;
input [3:0] f2dr_instpipe1, f2dr_instpipe2, f2dr_instpipe3;
input clock, flush, reset;
input [63:0] w2re_datapipe1, w2re_datapipe2, w2re_datapipe3;
input w2r_wrpipe1, w2r_wrpipe2, w2r_wrpipe3;
input [3:0] w2re_destpipe1, w2re_destpipe2, w2re_destpipe3;

output [63:0] r2e_src1datapipe1, r2e_src1datapipe2, r2e_src1datapipe3;
output [63:0] r2e_src2datapipe1, r2e_src2datapipe2, r2e_src2datapipe3;
output [3:0] r2e_src1pipe1, r2e_src1pipe2, r2e_src1pipe3;
output [3:0] r2e_src2pipe1, r2e_src2pipe2, r2e_src2pipe3;
```

TABLE 3.8   Interface Signals of _register file_ Module

Signal Name	Input/ Output	Bits	Description
clock	Input	1	Input clock pin. The VLIW microprocessor is active on rising edge of clock.
reset	Input	1	Input reset pin. Reset is asynchronous and active high.
flush	Input	1	This is a global signal that flushes all the modules, indicating that a branch is to occur.
f2r_src1pipe2	Input	4	Represents the source1 register for operation 2.
f2r_src1pipe3	Input	4	Represents the source1 register for operation 3.
f2r_src2pipe1	Input	4	Represents the source2 register for operation 1.
f2r_src2pipe2	Input	4	Represents the source2 register for operation 2.
f2r_src2pipe3	Input	4	Represents the source2 register for operation 3.
f2dr_instpipe1	Input	4	Represents the instruction of operation 1.
f2dr_instpipe2	Input	4	Represents the instruction of operation 2.
f2dr_instpipe3	Input	4	Represents the instruction of operation 3.
w2re_datapipe1	Input	64	Represents the 64-bit result of operation 1 executed by execute module. These data are written into the register file module if signal w2r_wrpipe1 is at logic 1.
w2re_datapipe2	Input	64	Represents the 64-bit result of operation 2 executed by execute module. These data are written into the register file module if signal w2r_wrpipe2 is at logic 1.
w2re_datapipe3	Input	64	Represents the 64-bit result of operation 3 executed by execute module. These data are written into the register file module if signal w2r_wrpipe3 is at logic 1.
w2r_wrpipe1	Input	1	Represents the write signal from writeback module to register file module. When this signal is logic 1, contents of w2re_datapipe1 is stored into register specified by w2re_destpipe1.
w2r_wrpipe2	Input	1	Represents the write signal from writeback module to register file module. When this signal is logic 1, contents of w2re_datapipe2 are stored into the register specified by w2re_destpipe2.
w2r_wrpipe3	Input	1	Represents the write signal from writeback module to register file module. When this signal is logic 1, contents of w2re_datapipe3 are stored into the register specified by w2re_destpipe3.
w2re_destpipe1	Input	4	Represents the destination register of operation 1.
w2re_destpipe2	Input	4	Represents the destination register of operation 2.
w2re_destpipe3	Input	4	Represents the destination register of operation 3.
r2e_src1datapipe1	Output	64	Represents the 64-bit data contents of register specified by r2e_src1pipe1. The contents are the source1 data of operation 1.
r2e_src1datapipe2	Output	64	Represents the 64-bit data contents of register specified by r2e_src1pipe2. The contents are the source1 data of operation 2.
r2e_src1datapipe3	Output	64	Represents the 64-bit data contents of register specified by r2e_src1pipe3. The contents are the source1 data of operation 3.

(Continued)

TABLE 3.8   Interface Signals of *register file* Module (*Continued*)

Signal Name	Input/ Output	Bits	Description
r2e_src2datapipe1	Output	64	Represents the 64-bit data contents of register specified by r2e_src2pipe1. The contents are the source2 data of operation 1.
r2e_src2datapipe2	Output	64	Represents the 64-bit data contents of register specified by r2e_src2pipe2. The contents are the source2 data of operation 2.
r2e_src2datapipe3	Output	64	Represents the 64-bit data contents of register specified by r2e_src2pipe3. The contents are the source2 data of operation 3.
r2e_src1pipe1	Output	4	Represents the source1 register of operation 1.
r2e_src1pipe2	Output	4	Represents the source1 register of operation 2.
r2e_src1pipe3	Output	4	Represents the source1 register of operation 3.
r2e_src2pipe1	Output	4	Represents the source2 register of operation 1.
r2e_src2pipe2	Output	4	Represents the source2 register of operation 2.
r2e_src2pipe3	Output	4	Represents the source2 register of operation 3.

```
reg [63:0] r2e_src1datapipe1, r2e_src1datapipe2, r2e_src1datapipe3;
reg [63:0] r2e_src2datapipe1, r2e_src2datapipe2, r2e_src2datapipe3;
reg [63:0] memoryarray [0:15];
reg [3:0] r2e_src1pipe1, r2e_src1pipe2, r2e_src1pipe3;
reg [3:0] r2e_src2pipe1, r2e_src2pipe2, r2e_src2pipe3;

integer i;

always @ (posedge clock or posedge reset)
begin
  if (reset)
```

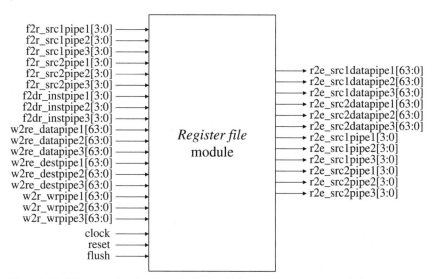

Figure 3.11 Diagram showing interface signals for register file module.

```
begin
  for (i=0; i < 16; i=i+1)
    memoryarray [i] <= 0;
  r2e_src1datapipe1 <= 0;
  r2e_src1datapipe2 <= 0;
  r2e_src1datapipe3 <= 0;
  r2e_src2datapipe1 <= 0;
  r2e_src2datapipe2 <= 0;
  r2e_src2datapipe3 <= 0;
  r2e_src1pipe1 <= 0;
  r2e_src1pipe2 <= 0;
  r2e_src1pipe3 <= 0;
  r2e_src2pipe1 <= 0;
  r2e_src2pipe2 <= 0;
  r2e_src2pipe3 <= 0;
end
else // positive edge clock detected
begin
  if (~flush) // not flushing
  begin
    // for instruction on pipe1
    case (f2dr_instpipe1)
    4'b0000:
    begin
      // nop, src registers are zero
      r2e_src1datapipe1 <= 0;
      r2e_src2datapipe1 <= 0;
    end
    4'b0001:
    begin
      // add inst
      r2e_src1datapipe1 <=
      memoryarray [f2r_src1pipe1];
      r2e_src2datapipe1 <=
      memoryarray [f2r_src2pipe1];
    end
    4'b0010:
    begin
      // sub inst
      r2e_src1datapipe1 <=
      memoryarray [f2r_src1pipe1];
      r2e_src2datapipe1 <=
      memoryarray [f2r_src2pipe1];
    end
    4'b0011:
    begin
      // mul inst
      // for mul, only bits 31 to 0 are
      // used. upper bits on src1 and
      // src2 are truncated to zeros
      r2e_src1datapipe1 <=
      64'h00000000ffffffff &
      memoryarray [f2r_src1pipe1];
      r2e_src2datapipe1 <=
      64'h00000000ffffffff &
      memoryarray [f2r_src2pipe1];
    end
    4'b0100:
    begin
      // load inst
      // register data from rf are all
      // zeros caused not needed
```

During reset, all the contents of register file are cleared to zero.

During nop, the source1 and source2 contents passed to execute module are zero.

In multiply operation, only lower 32 bits are used to form a 64-bit multiply result.

```
        r2e_src1datapipe1 <= 0;
        r2e_src2datapipe1 <= 0;
    end
4'b0101:
    begin
        // move inst
        // move src1 to dest
        r2e_src1datapipe1 <=
        memoryarray [f2r_src1pipe1];
        r2e_src2datapipe1 <= 0;
    end
4'b0110:
    begin
        // read inst - read src1
        r2e_src1datapipe1 <=
        memoryarray [f2r_src1pipe1];
        r2e_src2datapipe1 <= 0;
    end
4'b0111:
    begin
        // compare inst
        r2e_src1datapipe1 <=
        memoryarray [f2r_src1pipe1];
        r2e_src2datapipe1 <=
        memoryarray [f2r_src2pipe1];
    end
4'b1000:
    begin
        // xor inst
        r2e_src1datapipe1 <=
        memoryarray [f2r_src1pipe1];
        r2e_src2datapipe1 <=
        memoryarray [f2r_src2pipe1];
    end
4'b1001:
    begin
        // nand inst
        r2e_src1datapipe1 <=
        memoryarray [f2r_src1pipe1];
        r2e_src2datapipe1 <=
        memoryarray [f2r_src2pipe1];
    end
4'b1010:
    begin
        // nor inst
        r2e_src1datapipe1 <=
        memoryarray [f2r_src1pipe1];
        r2e_src2datapipe1 <=
        memoryarray [f2r_src2pipe1];
    end
4'b1011:
    begin
        // not inst
        // src2 are put to zeros since not
        // used
        r2e_src1datapipe1 <=
        memoryarray [f2r_src1pipe1];
        r2e_src2datapipe1 <= 0;
    end
```

> If a write and read occur simultaneously to the register file, the data read out at r2e_src1datapipe and r2e_src2datapipe will be stale data. This can be overcome by using the following code which uses more logic.
>
> ```
> r2e_src1datapipe1 <=
> (w2r_wrpipe1 & (w2re_destpipe1
> == f2r_src1pipe1)) ?
> w2re_datapipe1 : (w2r_wrpipe2
> & (w2re_destpipe2 ==
> f2r_src1pipe1)) ?
> w2re_datapipe2 : (w2r_wrpipe3
> & (w2re_destpipe3 ==
> f2r_src1pipe1)) ?
> w2re_datapipe3 : memoryarray
> [f2r_src1pipe1];
>
> r2e_src2datapipe1 <=
> (w2r_wrpipe1 & (w2re_destpipe1
> == f2r_src2pipe1)) ?
> w2re_datapipe1 : (w2r_wrpipe2
> & (w2re_destpipe2 ==
> f2r_src2pipe1)) ?
> w2re_datapipe2 : (w2r_wrpipe3
> & (w2re_destpipe3 ==
> f2r_src2pipe1)) ?
> w2re_datapipe3 : memoryarray
> [f2r_src2pipe1];
> ```
>
> Similarly code changes for pipe2 and pipe3.

```
   4'b1100:
     begin
       // shift left inst
       // src1 data shifted left
       // the amount of shift left decided
       // by src2[3:0]
       r2e_src1datapipe1 <=
       memoryarray [f2r_src1pipe1];
       r2e_src2datapipe1 <=
       64'h000000000000000f &
       memoryarray [f2r_src2pipe1];
     end
   4'b1101:
     begin
       // shift right inst
       // src1 data shifted right
       // the amount of shift right
       // decided by src2[3:0]
       r2e_src1datapipe1 <=
       memoryarray [f2r_src1pipe1];
       r2e_src2datapipe1 <=
       64'h000000000000000f &
       memoryarray [f2r_src2pipe1];
     end
   4'b1110:
     begin
       // barrel shift left inst
       // src1 data barrel shifted left
       // the amount of barrel shift left
       // decided by src2[3:0]
       r2e_src1datapipe1 <=
       memoryarray [f2r_src1pipe1];
       r2e_src2datapipe1 <=
       64'h000000000000000f &
       memoryarray [f2r_src2pipe1];
     end
   4'b1111:
     begin
       // barrel shift right inst
       // src1 data barrel shifted right
       // the amount of barrel shift right
       // decided by src2[3:0]
       r2e_src1datapipe1 <=
       memoryarray [f2r_src1pipe1];
       r2e_src2datapipe1 <=
       64'h000000000000000f &
       memoryarray [f2r_src2pipe1];
     end
   default:
     begin
       r2e_src1datapipe1 <= 0;
       r2e_src2datapipe1 <= 0;
     end
endcase
// for instruction on pipe2
case (f2dr_instpipe2)
   4'b0000:
     begin
       // nop, src registers are zero
       r2e_src1datapipe2 <= 0;
       r2e_src2datapipe2 <= 0;
     end
```

```
4'b0001:
  begin
    // add inst
    r2e_src1datapipe2 <=
    memoryarray [f2r_src1pipe2];
    r2e_src2datapipe2 <=
    memoryarray [f2r_src2pipe2];
  end
4'b0010:
  begin
    // sub inst
    r2e_src1datapipe2 <=
    memoryarray [f2r_src1pipe2];
    r2e_src2datapipe2 <=
    memoryarray [f2r_src2pipe2];
  end
4'b0011:
  begin
    // mul inst
    // for mul, only bits 31 to 0 are
    // used. upper bits on src1 and
    // src2 are truncated to zeros
    r2e_src1datapipe2 <=
    64'h00000000ffffffff & memoryarray
    [f2r_src1pipe2];
    r2e_src2datapipe2 <=
    64'h00000000ffffffff & memoryarray
    [f2r_src2pipe2];
  end
4'b0100:
  begin
    // load inst
    // register data from rf are all
    // zeros caused not needed
    r2e_src1datapipe2 <= 0;
    r2e_src2datapipe2 <= 0;
  end
4'b0101:
  begin
    // move inst
    // move src1 to dest
    r2e_src1datapipe2 <=
    memoryarray [f2r_src1pipe2];
    r2e_src2datapipe2 <= 0;
  end
4'b0110:
  begin
    // read inst - read src1
    r2e_src1datapipe2 <=
    memoryarray [f2r_src1pipe2];
    r2e_src2datapipe2 <= 0;
  end
4'b0111:
  begin
    // compare inst
    r2e_src1datapipe2 <=
    memoryarray [f2r_src1pipe2];
    r2e_src2datapipe2 <=
    memoryarray [f2r_src2pipe2];
  end
```

```
4'b1000:
  begin
    // xor inst
    r2e_src1datapipe2 <=
    memoryarray [f2r_src1pipe2];
    r2e_src2datapipe2 <=
    memoryarray [f2r_src2pipe2];
  end
4'b1001:
  begin
    // nand inst
    r2e_src1datapipe2 <=
    memoryarray [f2r_src1pipe2];
    r2e_src2datapipe2 <=
    memoryarray [f2r_src2pipe2];
  end
4'b1010:
  begin
    // nor inst
    r2e_src1datapipe2 <=
    memoryarray [f2r_src1pipe2];
    r2e_src2datapipe2 <=
    memoryarray [f2r_src2pipe2];
  end
4'b1011:
  begin
    // not inst
    // src2 are put to zeros since not
    // used
    r2e_src1datapipe2 <=
    memoryarray [f2r_src1pipe2];
    r2e_src2datapipe2 <= 0;
  end
4'b1100:
  begin
    // shift left inst
    // src1 data shifted left
    // the amount of shift left decided
    // by src2[3:0]
    r2e_src1datapipe2 <=
    memoryarray [f2r_src1pipe2];
    r2e_src2datapipe2 <=
    64'h000000000000000f &
    memoryarray [f2r_src2pipe2];
  end
4'b1101:
  begin
    // shift right inst
    // src1 data shifted right
    // the amount of shift right
    // decided by src2[3:0]
    r2e_src1datapipe2 <=
    memoryarray [f2r_src1pipe2];
    r2e_src2datapipe2 <=
    64'h000000000000000f &
    memoryarray [f2r_src2pipe2];
  end
4'b1110:
  begin
    // barrel shift left inst
    // src1 data barrel shifted left
    // the amount of barrel shift left
```

```
      // decided by src2[3:0]
      r2e_src1datapipe2 <=
      memoryarray [f2r_src1pipe2];
      r2e_src2datapipe2 <=
      64'h000000000000000f &
      memoryarray [f2r_src2pipe2];
    end
  4'b1111:
    begin
      // barrel shift right inst
      // src1 data barrel shifted right
      // the amount of barrel shift right
      // decided by src2[3:0]
      r2e_src1datapipe2 <=
      memoryarray [f2r_src1pipe2];
      r2e_src2datapipe2 <=
      64'h000000000000000f &
      memoryarray [f2r_src2pipe2];
    end
  default:
    begin
      r2e_src1datapipe2 <= 0;
      r2e_src2datapipe2 <= 0;
    end
endcase

// for instruction on pipe3
case (f2dr_instpipe3)
  4'b0000:
    begin
      // nop, src registers are zero
      r2e_src1datapipe3 <= 0;
      r2e_src2datapipe3 <= 0;
    end
  4'b0001:
    begin
      // add inst
      r2e_src1datapipe3 <=
      memoryarray [f2r_src1pipe3];
      r2e_src2datapipe3 <=
      memoryarray [f2r_src2pipe3];
    end
  4'b0010:
    begin
      // sub inst
      r2e_src1datapipe3 <=
      memoryarray [f2r_src1pipe3];
      r2e_src2datapipe3 <=
      memoryarray [f2r_src2pipe3];
    end
  4'b0011:
    begin
      // mul inst
      // for mul, only bits 31 to 0 are
      // used
      // upper bits on src1 and src2 are
      // truncated to zeros
      r2e_src1datapipe3 <=
      64'h00000000ffffffff &
      memoryarray [f2r_src1pipe3];
      r2e_src2datapipe3 <=
```

```verilog
        64'h00000000ffffffff &
        memoryarray [f2r_src2pipe3];
    end
4'b0100:
  begin
    // load inst
    // register data from rf are all
    // zeros caused not needed
    r2e_src1datapipe3 <= 0;
    r2e_src2datapipe3 <= 0;
  end
4'b0101:
  begin
    // move inst
    // move src1 to dest
    r2e_src1datapipe3 <=
    memoryarray [f2r_src1pipe3];
    r2e_src2datapipe3 <= 0;
  end
4'b0110:
  begin
    // read inst - read src1
    r2e_src1datapipe3 <=
    memoryarray [f2r_src1pipe3];
    r2e_src2datapipe3 <= 0;
  end
4'b0111:
  begin
    // compare inst
    r2e_src1datapipe3 <=
    memoryarray [f2r_src1pipe3];
    r2e_src2datapipe3 <=
    memoryarray [f2r_src2pipe3];
  end
4'b1000:
  begin
    // xor inst
    r2e_src1datapipe3 <=
    memoryarray [f2r_src1pipe3];
    r2e_src2datapipe3 <=
    memoryarray [f2r_src2pipe3];
  end
4'b1001:
  begin
    // nand inst
    r2e_src1datapipe3 <=
    memoryarray [f2r_src1pipe3];
    r2e_src2datapipe3 <=
    memoryarray [f2r_src2pipe3];
  end
4'b1010:
  begin
    // nor inst
    r2e_src1datapipe3 <=
    memoryarray [f2r_src1pipe3];
    r2e_src2datapipe3 <=
    memoryarray [f2r_src2pipe3];
  end
4'b1011:
  begin
    // not inst
    // src2 are put to zeros since not
```

```verilog
    // used
    r2e_src1datapipe3 <=
    memoryarray [f2r_src1pipe3];
    r2e_src2datapipe3 <= 0;
   end
4'b1100:
  begin
    // shift left inst
    // src1 data shifted left
    // the amount of shift left decided
    // by src2[3:0]
    r2e_src1datapipe3 <=
    memoryarray [f2r_src1pipe3];
    r2e_src2datapipe3 <=
    64'h000000000000000f &
    memoryarray [f2r_src2pipe3];
   end
4'b1101:
  begin
    // shift right inst
    // src1 data shifted right
    // the amount of shift right
    // decided by src2[3:0]
    r2e_src1datapipe3 <=
    memoryarray [f2r_src1pipe3];
    r2e_src2datapipe3 <=
    64'h000000000000000f &
    memoryarray [f2r_src2pipe3];
   end
4'b1110:
  begin
    // barrel shift left inst
    // src1 data barrel shifted left
    // the amount of barrel shift left
    // decided by src2[3:0]
    r2e_src1datapipe3 <=
    memoryarray [f2r_src1pipe3];
    r2e_src2datapipe3 <=
    64'h000000000000000f &
    memoryarray [f2r_src2pipe3];
   end
4'b1111:
  begin
    // barrel shift right inst
    // src1 data barrel shifted right
    // the amount of barrel shift right
    // decided by src2[3:0]
    r2e_src1datapipe3 <=
    memoryarray [f2r_src1pipe3];
    r2e_src2datapipe3 <=
    64'h000000000000000f &
    memoryarray [f2r_src2pipe3];
   end
default:
  begin
    r2e_src1datapipe3 <= 0;
    r2e_src2datapipe3 <= 0;
   end
 endcase
r2e_src1pipe1 <= f2r_src1pipe1;
r2e_src1pipe2 <= f2r_src1pipe2;
r2e_src1pipe3 <= f2r_src1pipe3;
```

```
    r2e_src2pipe1 <= f2r_src2pipe1;
    r2e_src2pipe2 <= f2r_src2pipe2;
    r2e_src2pipe3 <= f2r_src2pipe3;
end
else // flush the pipe
begin
    r2e_src1datapipe1 <= 0;
    r2e_src1datapipe2 <= 0;
    r2e_src1datapipe3 <= 0;
    r2e_src2datapipe1 <= 0;
    r2e_src2datapipe2 <= 0;
    r2e_src2datapipe3 <= 0;
    r2e_src1pipe1 <= 0;
    r2e_src1pipe2 <= 0;
    r2e_src1pipe3 <= 0;
    r2e_src2pipe1 <= 0;
    r2e_src2pipe2 <= 0;
    r2e_src2pipe3 <= 0;
end

// writing of data into register file for pipe1
if (w2r_wrpipe1)
  memoryarray [w2re_destpipe1]
   <= w2re_datapipe1;

// writing of data into register file
for pipe2
if (w2r_wrpipe2)
  memoryarray [w2re_destpipe2] <= w2re_datapipe2;

// writing of data into register file for pipe3
if (w2r_wrpipe3)
  memoryarray [w2re_destpipe3] <= w2re_datapipe3;
  end
end
endmodule
```

Writing of contents into the register file module

Referring to the RTL code of register file module shown in Example 3.8, there are two possible scenarios that can "break" the design:

1. A possible data corruption situation may occur. If the write and read operations to the register file module occur at the same time to the same register location, data corruption may happen. For example, if signal w2r_wrpipe1 is at logic 1, w2re_destpipe1 is reg5, f2dr_instpipe1 is 0001 (add operation), f2r_src1pipe1 is reg5, a read and write both occur at reg5 at the same time. This situation will lead to the data at bus r2e_src1datapipe1 potentially being corrupted, causing the execute module to execute on corrupted data.

2. A possible stale data situation may occur when there is data dependency between different VLIW instructions. Consider the following two VLIW instructions:

```
load #0001, r1 : load #0002, r2 : load #0003, r3
    add r1, r2, r4 : sub r2, r3, r5 : read r6
```

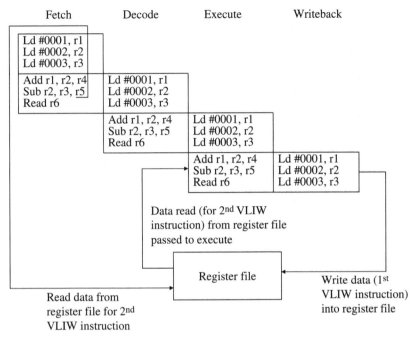

Figure 3.12 Diagram showing two VLIW instructions passing through the VLIW microprocessor four-stage pipeline.

The first VLIW operation consists of three load operations to load data into register reg1, reg2, and reg3. The second VLIW operation consists of an addition function of contents reg1 and reg2, a subtraction function of contents of reg2 and reg3, and a read function of reg6. Figure 3.12 shows the two VLIW instructions passing through the four-stage pipeline of the VLIW microprocessor.

Referring to Figure 3.12, at the writeback stage, the data from the load operation are written into the register file at register reg1, reg2, and reg3. However, before the writeback stage can complete writing the data into the three mentioned registers, the contents of register reg1, reg2, and reg3 are read. The contents of reg1, reg2, and reg3 are passed to the execute stage to allow the execution of the operations of the second VLIW instruction (add r1,r2,r4:sub r2,r3,r5:read r6). This creates a situation of stale data being passed to the execute stage as the data from the load operation have not been written into the registers.

To workaround the problem of data corruption and stale data, a design concept called register bypassing is introduced to the VLIW microprocessor. Register bypassing allows the data to be written into the

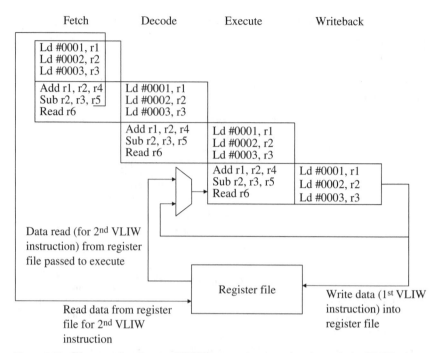

Figure 3.13 Diagram showing two VLIW instructions passing through the VLIW microprocessor four-stage pipeline with register bypassing.

register file to be bypassed back into the execute stage. This allows the data to be written into the register file to be used directly by the execute stage instead of reading stale data from the register file. Register bypassing also eliminates the possibility of data at bus r2e_src1datapipe1, r2e_src1datapipe2, r2e_src1datapipe3 and r2e_src2datapipe1, r2e_src2datapipe2, r2e_src2datapipe3 being corrupted. From an architectural perspective, implementation of register bypassing is shown in Figure 3.13.

In Figure 3.13, the data output of the writeback stage is bypassed back into the execute stage through a multiplexer, while at the same time being written into the corresponding register in the register file. The multiplexer chooses between the contents read from the register file or the output of the writeback stage to be used by the execute stage. Which data are chosen depends on whether instruction dependency is detected by the VLIW microprocessor. This will require additional logic to monitor for instruction dependency.

For the VLIW microprocessor, the feature of register bypass and instruction dependency monitoring is implemented in the execute stage, which is discussed in Section 3.2.4.

> *Note*: For a detailed explanation on register bypass, please refer to
> *Computer Architecture, Third Edition: A Quantitative Approach* by John L.
> Hennessy and David A. Patterson (Morgan Kaufmann).

3.2.4 Module *execute* RTL Code

The execute module is the most complex and complicated module
within the VLIW microprocessor. Its functionality is to execute the
operations of the VLIW instruction. The feature of register bypassing
is also implemented in the execute module.

Table 3.9 shows the interface signals for the execute module and its
interface signal functionality. Figure 3.14 shows the interface signal
diagram of the execute module.

TABLE 3.9 Interface Signals of execute Module

Signal Name	Input/ Output	Bits	Description
clock	Input	1	Input clock pin. The VLIW micro-processor is active on rising edge of clock.
reset	Input	1	Input reset pin. Reset is asynchronous and active high.
d2e_destpipe1	Input	4	Represents the destination register for operation 1.
d2e_destpipe2	Input	4	Represents the destination register for operation 2.
d2e_destpipe3	Input	4	Represents the destination register for operation 3.
d2e_instpipe1	Input	4	Represents the instruction of operation 1.
d2e_instpipe2	Input	4	Represents the instruction of operation 2.
d2e_instpipe3	Input	4	Represents the instruction of operation 3.
d2e_datapipe1	Input	64	Represents the data for operation 1. The data bus is used only for load instruction.
d2e_datapipe2	Input	64	Represents the data for operation 2. The data bus is used only for load instruction.
d2e_datapipe3	Input	64	Represents the data for operation 3. The data bus is used only for load instruction.
r2e_src1datapipe1	Input	64	Represents the 64-bit data contents of register specified by r2e_src1pipe1. The contents are the source1 data of operation 1.
r2e_src1datapipe2	Input	64	Represents the 64-bit data contents of register specified by r2e_src1pipe2. The contents are the source1 data of operation 2.
r2e_src1datapipe3	Input	64	Represents the 64-bit data contents of register specified by r2e_src1pipe3. The contents are the source1 data of operation 3.

(Continued)

TABLE 3.9   Interface Signals of *execute* Module (*Continued*)

Signal Name	Input/Output	Bits	Description
r2e_src2datapipe1	Input	64	Represents the 64-bit data contents of register specified by r2e_src2pipe1. The contents are the source2 data of operation 1.
r2e_src2datapipe2	Input	64	Represents the 64-bit data contents of register specified by r2e_src2pipe2. The contents are the source2 data of operation 2.
r2e_src2datapipe3	Input	64	Represents the 64-bit data contents of register specified by r2e_src2pipe3. The contents are the source2 data of operation 3.
r2e_src1pipe1	Input	4	Represents the source1 register of operation 1.
r2e_src1pipe2	Input	4	Represents the source1 register of operation 2.
r2e_src1pipe3	Input	4	Represents the source1 register of operation 3.
r2e_src2pipe1	Input	4	Represents the source2 register of operation 1.
r2e_src2pipe2	Input	4	Represents the source2 register of operation 2.
r2e_src2pipe3	Input	4	Represents the source2 register of operation 3.
w2re_destpipe1	Input	4	Represents the destination register of operation 1.
w2re_destpipe2	Input	4	Represents the destination register of operation 2.
w2re_destpipe3	Input	4	Represents the destination register of operation 3.
w2re_datapipe1	Input	64	Represents the 64-bit result of operation 1.
w2re_datapipe2	Input	64	Represents the 64-bit result of operation 2.
w2re_datapipe3	Input	64	Represents the 64-bit result of operation 3.
flush	Output	1	This is a global signal that flushes all the modules, indicating that a branch is to occur.
e2w_destpipe1	Output	4	Represents the destination register for operation 1.
e2w_destpipe2	Output	4	Represents the destination register for operation 2.
e2w_destpipe3	Output	4	Represents the destination register for operation 3.
e2w_datapipe1	Output	64	Represents the data for operation 1.
e2w_datapipe2	Output	64	Represents the data for operation 2.
e2w_datapipe3	Output	64	Represents the data for operation 3.

(Continued)

TABLE 3.9   Interface Signals of *execute* Module (*Continued*)

Signal Name	Input/ Output	Bits	Description
e2w_wrpipe1	Output	1	Represents the write signal from execute module to writeback module. This signal is passed from writeback module to register file module. It indicates the contents of w2re_datapipe1 to be stored into register specified by w2re_destpipe1.
e2w_wrpipe2	Output	1	Represents the write signal from execute module to writeback module. This signal is passed from writeback module to register file module. It indicates the contents of w2re_datapipe2 to be stored into register specified by w2re_destpipe2.
e2w_wrpipe3	Output	1	Represents the write signal from execute module to writeback module. This signal is passed from writeback module to register file module. It indicates the contents of w2re_datapipe3 to be stored into register specified by w2re_destpipe3.
e2w_readpipe1	Output	1	This signal indicates to the writeback module that the data on e2w_datapipe1 are to be read out of the VLIW microprocessor, through the output port readdatapipe1.
e2w_readpipe2	Output	1	This signal indicates to the writeback module that the data on e2w_datapipe2 are to be read out of the VLIW microprocessor, through the output port readdatapipe2.
e2w_readpipe3	Output	1	This signal indicates to the writeback module that the data on e2w_datapipe3 are to be read out of the VLIW microprocessor, through the output port readdatapipe3.
jump	Output	1	This signal indicates to the external instruction memory module that a branch to another VLIW instruction is occurring. The external instruction memory module will fetch the newly branched instruction.

Based on the interface signals shown in Table 3.9 with the signal functionality, the RTL verilog code for the execute module is shown in Example 3.9. A significant portion of the logic required for the execute module is for the register bypassing mechanism for avoiding data corruption and stale data as discussed in Section 3.2.3.

For the implementation of a VLIW microprocessor which consists of three separate operations executed in parallel, the register bypass logic

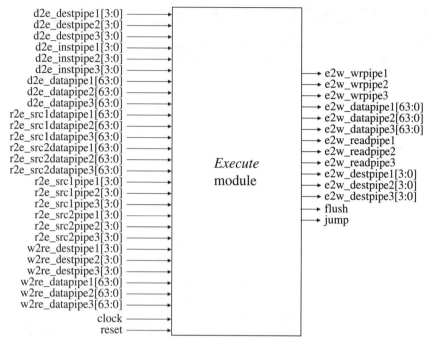

Figure 3.14 Diagram showing interface signals for execute module.

must be able to cater for interpipe and intrapipe data dependency within the VLIW microprocessor.

Among the different conditions which require register bypassing are the following:

1. operation1 to operation1 bypass on source1

   A register bypass is required when two VLIW instructions with operation1 on each instruction referencing to the same register on source1.

```
>ld #fafa, r0: ld #abab, r1: ld #bcbc, r2
ld #1234, r3: ld #5678, r4: ld #9abc, r5
add r3, r0, r10: sub r0, r1, r11: mul r1, r2, r12
```

fetch module	decode module	execute module	writeback module
ld #fafa, r0			
ld #abab, r1			
ld #bcbc, r2			
ld #1234, r3	ld #fafa, r0		
ld #5678, r4	ld #abab, r1		
ld #9abc, r5	ld #bcbc, r2		

add r3,r0,r10	ld #1234, r3	ld #fafa, r0	
sub r0,r1,r11	ld #5678, r4	ld #abab, r1	
mul r1,r2,r12	ld #9abc, r5	ld #bcbc, r2	
	add r3,r0,r10	ld #1234, r3	ld #fafa, r0
	sub r0,r1,r11	ld #5678, r4	ld #abab, r1
	mul r1,r2,r12	ld #9abc, r5	ld #bcbc, r2
		add r3,r0,r10	ld #1234, r3
		sub r0,r1,r11	ld #5678, r4
		mul r1,r2,r12	ld #9abc, r5
			add r3,r0,r10
			sub r0,r1,r11
			mul r1,r2,r12

When the VLIW instruction ld #1234, r3: ld #5678, r4: ld #9abc, r5 reaches the writeback module, the VLIW instruction add r3, r0, r10: sub r0, r1, r11: mul r1, r2, r12 reaches the execute module. Before the data #1234 is written into register r3 in the register file, the contents of register r3 are read and passed from the register file to the execute module for use on the VLIW instruction add r3, r0, r10: sub r0, r1, r11: mul r1, r2, r12. This causes stale data on the contents of register r3. To work around this problem, a register bypass is required between source1 data of operation1 to operation1 of VLIW instruction.

2. operation1 to operation1 bypass on source2

3. operation1 to operation2 bypass on source1

4. operation1 to operation2 bypass on source2

5. operation1 to operation3 bypass on source1

6. operation1 to operation3 bypass on source2

7. operation1 to operation1 bypass on source1, and operation1 to operation1 bypass on source2

8. operation1 to operation1 bypass on source1, and operation1 to operation2 bypass on source2

9. operation1 to operation1 bypass on source1, and operation1 to operation3 bypass on source2

10. operation1 to operation1 bypass on source1, and operation2 to operation1 bypass on source2

11. operation1 to operation1 bypass on source1, and operation2 to operation2 bypass on source2

12. operation1 to operation1 bypass on source1, and operation2 to operation3 bypass on source2

13. operation1 to operation1 bypass on source1, and operation3 to operation1 bypass on source2

14. operation1 to operation1 bypass on source1, and operation3 to operation2 bypass on source2

15. operation1 to operation1 bypass on source1, and operation3 to operation3 bypass on source2

16. operation1 to operation2 bypass on source1, and operation1 to operation1 bypass on source2

17. operation1 to operation2 bypass on source1, and operation1 to operation2 bypass on source2

18. operation1 to operation2 bypass on source1, and operation1 to operation3 bypass on source2

19. operation1 to operation2 bypass on source1, and operation2 to operation1 bypass on source2

20. operation1 to operation2 bypass on source1, and operation2 to operation2 bypass on source2

21. operation1 to operation2 bypass on source1, and operation2 to operation3 bypass on source2

22. operation1 to operation2 bypass on source1, and operation3 to operation1 bypass on source2

23. operation1 to operation2 bypass on source1, and operation3 to operation2 bypass on source2

24. operation1 to operation2 bypass on source1, and operation3 to operation3 bypass on source2

25. operation1 to operation3 bypass on source1, and operation1 to operation1 bypass on source2

26. operation1 to operation3 bypass on source1, and operation1 to operation2 bypass on source2

27. operation1 to operation3 bypass on source1, and operation1 to operation3 bypass on source2

28. operation1 to operation3 bypass on source1, and operation2 to operation1 bypass on source2

29. operation1 to operation3 bypass on source1, and operation2 to operation2 bypass on source2

30. operation1 to operation3 bypass on source1, and operation2 to operation3 bypass on source2

31. operation1 to operation3 bypass on source1, and operation3 to operation1 bypass on source2

32. operation1 to operation3 bypass on source1, and operation3 to operation2 bypass on source2

33. operation1 to operation3 bypass on source1, and operation3 to operation3 bypass on source2

34. operation2 to operation1 bypass on source1

35. operation2 to operation1 bypass on source2

36. operation2 to operation2 bypass on source1

37. operation2 to operation2 bypass on source2

38. operation2 to operation3 bypass on source1

39. operation2 to operation3 bypass on source2

40. operation2 to operation1 bypass on source1, and operation1 to operation1 bypass on source2

41. operation2 to operation1 bypass on source1, and operation1 to operation2 bypass on source2

42. operation2 to operation1 bypass on source1, and operation1 to operation3 bypass on source2

43. operation2 to operation1 bypass on source1, and operation2 to operation1 bypass on source2

44. operation2 to operation1 bypass on source1, and operation2 to operation2 bypass on source2

45. operation2 to operation1 bypass on source1, and operation2 to operation3 bypass on source2

46. operation2 to operation1 bypass on source1, and operation3 to operation1 bypass on source2

47. operation2 to operation1 bypass on source1, and operation3 to operation2 bypass on source2

48. operation2 to operation1 bypass on source1, and operation3 to operation3 bypass on source2

49. operation2 to operation2 bypass on source1, and operation1 to operation1 bypass on source2

50. operation2 to operation2 bypass on source1, and operation1 to operation2 bypass on source2

51. operation2 to operation2 bypass on source1, and operation1 to operation3 bypass on source2

52. operation2 to operation2 bypass on source1, and operation2 to operation1 bypass on source2

53. operation2 to operation2 bypass on source1, and operation2 to operation2 bypass on source2

54. operation2 to operation2 bypass on source1, and operation2 to operation3 bypass on source2

55. operation2 to operation2 bypass on source1, and operation3 to operation1 bypass on source2

56. operation2 to operation2 bypass on source1, and operation3 to operation2 bypass on source2

57. operation2 to operation2 bypass on source1, and operation3 to operation3 bypass on source2

58. operation2 to operation3 bypass on source1, and operation1 to operation1 bypass on source2

59. operation2 to operation3 bypass on source1, and operation1 to operation2 bypass on source2

60. operation2 to operation3 bypass on source1, and operation1 to operation3 bypass on source2

61. operation2 to operation3 bypass on source1, and operation2 to operation1 bypass on source2

62. operation2 to operation3 bypass on source1, and operation2 to operation2 bypass on source2

63. operation2 to operation3 bypass on source1, and operation2 to operation3 bypass on source2

64. operation2 to operation3 bypass on source1, and operation3 to operation1 bypass on source2

65. operation2 to operation3 bypass on source1, and operation3 to operation2 bypass on source2

66. operation2 to operation3 bypass on source1, and operation3 to operation3 bypass on source2

67. operation3 to operation1 bypass on source1

68. operation3 to operation1 bypass on source2

69. operation3 to operation2 bypass on source1

70. operation3 to operation2 bypass on source2

71. operation3 to operation3 bypass on source1

72. operation3 to operation3 bypass on source2

73. operation3 to operation1 bypass on source1, and operation1 to operation1 bypass on source2

74. operation3 to operation1 bypass on source1, and operation1 to operation2 bypass on source2

75. operation3 to operation1 bypass on source1, and operation1 to operation3 bypass on source2

76. operation3 to operation1 bypass on source1, and operation2 to operation1 bypass on source2

77. operation3 to operation1 bypass on source1, and operation2 to operation2 bypass on source2

78. operation3 to operation1 bypass on source1, and operation2 to operation3 bypass on source2

79. operation3 to operation1 bypass on source1, and operation3 to operation1 bypass on source2

80. operation3 to operation1 bypass on source1, and operation3 to operation2 bypass on source2

81. operation3 to operation1 bypass on source1, and operation3 to operation3 bypass on source2

82. operation3 to operation2 bypass on source1, and operation1 to operation1 bypass on source2

83. operation3 to operation2 bypass on source1, and operation1 to operation2 bypass on source2

84. operation3 to operation2 bypass on source1, and operation1 to operation3 bypass on source2

85. operation3 to operation2 bypass on source1, and operation2 to operation1 bypass on source2

86. operation3 to operation2 bypass on source1, and operation2 to operation2 bypass on source2

87. operation3 to operation2 bypass on source1, and operation2 to operation3 bypass on source2

88. operation3 to operation2 bypass on source1, and operation3 to operation1 bypass on source2

89. operation3 to operation2 bypass on source1, and operation3 to operation2 bypass on source2

90. operation3 to operation2 bypass on source1, and operation3 to operation3 bypass on source2

91. operation3 to operation3 bypass on source1, and operation1 to operation1 bypass on source2

92. operation3 to operation3 bypass on source1, and operation1 to operation2 bypass on source2

93. operation3 to operation3 bypass on source1, and operation1 to operation3 bypass on source2

94. operation3 to operation3 bypass on source1, and operation2 to operation1 bypass on source2

95. operation3 to operation3 bypass on source1, and operation2 to operation2 bypass on source2

96. operation3 to operation3 bypass on source1, and operation2 to operation3 bypass on source2

97. operation3 to operation3 bypass on source1, and operation3 to operation1 bypass on source2

98. operation3 to operation3 bypass on source1, and operation3 to operation2 bypass on source2

99. operation3 to operation3 bypass on source1, and operation3 to operation3 bypass on source2

There are a total of 99 different conditions that require register bypass mechanism, as register bypassing is required not only for intraoperation (between operation1 to operation1 or operation2 to operation2 or operation3 to operation3) but also for interoperation (between operation1 and operation2, operation2 to operation3 and so forth). As such, the amount of logic required in the execute module for register bypass is rather significant and complex. This complexity increases when the VLIW microprocessor increases the number of operations in parallel. For example, a VLIW microprocessor that has four operations combined into one VLIW instruction increases the register bypassing logic significantly. For four parallel pipes, a total of 288 different conditions require register bypassing. Example 3.9 shows the RTL code for the execute module, with logic for register bypassing for all mentioned 99 different conditions.

Note: To ease understanding on the RTL code of the execute module, only partial bypassing logic is implemented. For load instructions, a written register value can only be used for instruction dependency after 2 clocks. For other instructions, a written register value can only be used for instruction dependency after 1 clock for intrapipe bypass and 2 clocks for interpipe bypass.

Example 3.9   RTL Verilog Code of execute Module

```
module execute (clock, reset,
d2e_instpipe1, d2e_instpipe2, d2e_instpipe3,
d2e_destpipe1, d2e_destpipe2, d2e_destpipe3,
d2e_datapipe1, d2e_datapipe2, d2e_datapipe3,
r2e_src1datapipe1, r2e_src1datapipe2, r2e_src1datapipe3,
r2e_src2datapipe1, r2e_src2datapipe2, r2e_src2datapipe3,
r2e_src1pipe1, r2e_src1pipe2, r2e_src1pipe3,
r2e_src2pipe1, r2e_src2pipe2, r2e_src2pipe3,
w2re_destpipe1, w2re_destpipe2, w2re_destpipe3,
w2re_datapipe1, w2re_datapipe2, w2re_datapipe3,
e2w_destpipe1, e2w_destpipe2, e2w_destpipe3,
e2w_datapipe1, e2w_datapipe2, e2w_datapipe3,
e2w_wrpipe1, e2w_wrpipe2, e2w_wrpipe3,
e2w_readpipe1, e2w_readpipe2, e2w_readpipe3,
flush, jump);
```

```
input clock, reset;
input [3:0] d2e_destpipe1, d2e_destpipe2, d2e_destpipe3;
input [3:0] d2e_instpipe1, d2e_instpipe2, d2e_instpipe3;
input [63:0] d2e_datapipe1, d2e_datapipe2, d2e_datapipe3;
input [63:0] r2e_src1datapipe1, r2e_src1datapipe2, r2e_src1datapipe3;
input [63:0] r2e_src2datapipe1, r2e_src2datapipe2, r2e_src2datapipe3;

input [3:0] r2e_src1pipe1, r2e_src1pipe2, r2e_src1pipe3;
input [3:0] r2e_src2pipe1, r2e_src2pipe2, r2e_src2pipe3;
input [3:0] w2re_destpipe1, w2re_destpipe2, w2re_destpipe3;
input [63:0] w2re_datapipe1, w2re_datapipe2, w2re_datapipe3;

output [3:0] e2w_destpipe1, e2w_destpipe2, e2w_destpipe3;
output [63:0] e2w_datapipe1, e2w_datapipe2, e2w_datapipe3;
output e2w_wrpipe1, e2w_wrpipe2, e2w_wrpipe3;
output e2w_readpipe1, e2w_readpipe2, e2w_readpipe3;
output flush, jump;

reg [3:0] e2w_destpipe1, e2w_destpipe2, e2w_destpipe3;
reg [63:0] e2w_datapipe1, e2w_datapipe2, e2w_datapipe3;
reg e2w_wrpipe1, e2w_wrpipe2, e2w_wrpipe3;
reg e2w_readpipe1, e2w_readpipe2, e2w_readpipe3;
reg flush, jump;
reg preflush;

reg [63:0] int_src1datapipe1, int_src1datapipe2, int_src1datapipe3;
reg [63:0] int_src2datapipe1, int_src2datapipe2, int_src2datapipe3;

reg [3:0] postw2re_destpipe1, postw2re_destpipe2, postw2re_destpipe3;
reg [63:0] postw2re_datapipe1, postw2re_datapipe2, postw2re_datapipe3;

// include the file that declares the parameter declaration for
// register names and also instruction operations
`include "regname.v"

always @ (posedge clock or posedge reset)
begin
  if (reset)
    begin
      postw2re_destpipe1 <= reg0;
      postw2re_datapipe1 <= 0;
      postw2re_destpipe2 <= reg0;
      postw2re_datapipe2 <= 0;
      postw2re_destpipe3 <= reg0;
      postw2re_datapipe3 <= 0;
    end
  else
    begin
      postw2re_destpipe1 <= w2re_destpipe1;
      postw2re_datapipe1 <= w2re_datapipe1;
      postw2re_destpipe2 <= w2re_destpipe2;
      postw2re_datapipe2 <= w2re_datapipe2;
      postw2re_destpipe3 <= w2re_destpipe3;
      postw2re_datapipe3 <= w2re_datapipe3;
    end
end
wire comp_w2re_dest = (w2re_destpipe1 == w2re_destpipe2)
  & (w2re_destpipe2 == w2re_destpipe3);
wire comp_postw2re_dest = (postw2re_destpipe1 == postw2re_destpipe2)
  & (postw2re_destpipe2 == postw2re_destpipe3);

// for register bypass for operation1
always @ (d2e_instpipe1 or postw2re_destpipe1 or r2e_src1pipe1 or
r2e_src2pipe1 or r2e_src1datapipe1 or r2e_src2datapipe1 or
postw2re_datapipe1 or w2re_destpipe1 or w2re_datapipe1 or
```

```verilog
e2w_wrpipe1 or postw2re_destpipe2 or postw2re_datapipe2 or
postw2re_destpipe3 or postw2re_datapipe3 or comp_w2re_dest or
comp_postw2re_dest)
begin
  if ((d2e_instpipe1 == load) | (d2e_instpipe1 == nop))
    begin
      int_src1datapipe1 = r2e_src1datapipe1;
      int_src2datapipe1 = r2e_src2datapipe1;
    end
  /* else if (e2w_wrpipe1) // for debug only
    begin
      if (postw2re_destpipe1 == r2e_src1pipe1)
        begin
          int_src1datapipe1 = postw2re_datapipe1;
          int_src2datapipe1 = r2e_src2datapipe1;
        end
      else if (postw2re_destpipe1 == r2e_src2pipe1)
        begin
          int_src1datapipe1 = r2e_src1datapipe1;
          int_src2datapipe1 = postw2re_datapipe1;
        end
      else
        begin
          int_src1datapipe1 = r2e_src1datapipe1;
          int_src2datapipe1 = r2e_src2datapipe1;
        end */
    end
  else
    begin
      if ((w2re_destpipe1 == r2e_src1pipe1)
        & ~comp_w2re_dest)
        begin
          int_src1datapipe1 = w2re_datapipe1;
          int_src2datapipe1 = r2e_src2datapipe1;
        end
      else if (((w2re_destpipe1 == r2e_src2pipe1)
        & ~((w2re_destpipe1 == reg0)
        &(r2e_src2pipe1==reg0)&(d2e_instpipe1
        ==read)) &~comp_w2re_dest)
        begin
          int_src1datapipe1 = r2e_src1datapipe1;
          int_src2datapipe1 = w2re_datapipe1;
        end
      // for cross operation register bypass between
      // operation3 and operation1 for src2 AND
      // between operation2 and operation1 for src1.
      else if ((postw2re_destpipe2 == r2e_src1pipe1) &
      (postw2re_destpipe3 == r2e_src2pipe1))
        begin
        case (d2e_instpipe1)
          4'b0011: // mul
            begin
              int_src1datapipe1 = 64'h00000000ffffffff
                & postw2re_datapipe2;
              int_src2datapipe1 = 64'h00000000ffffffff
                & postw2re_datapipe3;
            end
          4'b1100:
            // shift left inst.
            begin
              int_src1datapipe1 = postw2re_datapipe2;
              int_src2datapipe1 =64'h000000000000000f
                & postw2re_datapipe3;
            end
```

> This portion of the code is for intrapipe bypass for w2re_destpipe. Extend the "else if (w2re_destpipe" portion of this code for interpipe bypass, similar to the code for the bypass for postw2re_dest pipe below.

> Shift left inst. src1 data shifted left. Amount of shift left decided by src2[3:0].

```
    4'b1101:
    // shift right inst.
    begin
      int_src1datapipe1 = postw2re_datapipe2;
      int_src2datapipe1 =64'h000000000000000f
        & postw2re_datapipe3;
    end
    4'b1110:
    // barrel shift left inst.
    begin
      int_src1datapipe1 = postw2re_datapipe2;
      int_src2datapipe1= 64'h000000000000000f
        & postw2re_datapipe3;
    end
    4'b1111:
    // barrel shift right inst.
    begin
      int_src1datapipe1 = postw2re_datapipe2;
      int_src2datapipe1 =64'h000000000000000f
        & postw2re_datapipe3;
    end
    default:
    begin
      int_src1datapipe1 = postw2re_datapipe2;
      int_src2datapipe1 = postw2re_datapipe3;
    end
  endcase
 end
// for cross operation register bypass between
// operation3 and operation1 for src1 AND
// between operation2 and operation1 for src2.
else if ((postw2re_destpipe2 == r2e_src2pipe1) &
  (postw2re_destpipe3 == r2e_src1pipe1))
 begin
   case (d2e_instpipe1)
     4'b0011: // mul
       begin
         int_src1datapipe1 = 64'h00000000ffffffff
           & postw2re_datapipe3;
         int_src2datapipe1 = 64'h00000000ffffffff
           & postw2re_datapipe2;
       end
     4'b1100: // shift left inst.
       begin
         int_src1datapipe1 = postw2re_datapipe3;
         int_src2datapipe1 =64'h000000000000000f
           & postw2re_datapipe2;
       end
     4'b1101: // shift right inst.
       begin
         int_src1datapipe1 = postw2re_datapipe3;
         int_src2datapipe1 =64'h000000000000000f
           & postw2re_datapipe2;
       end
     4'b1110: // barrel shift left inst.
       begin
         int_src1datapipe1 = postw2re_datapipe3;
         int_src2datapipe1 =64'h000000000000000f
           & postw2re_datapipe2;
       end
```

> Shift right inst. src1 data shifted right. Amount of shift left decided by src2[3:0].

```verilog
4'b1111:  // barrel shift right inst.
  begin
    int_src1datapipe1 = postw2re_datapipe3;
    int_src2datapipe1 =64'h000000000000000f
      & postw2re_datapipe2;
  end
default:
  begin
    int_src1datapipe1 = postw2re_datapipe3;
    int_src2datapipe1 = postw2re_datapipe2;
  end
endcase
end
// for cross operation register bypass between
// operation1 and operation1 for src2 AND
// between operation3 and operation1 for src1.
else if ((postw2re_destpipe1 == r2e_src2pipe1) &
  (postw2re_destpipe3 == r2e_src1pipe1))
begin
  case (d2e_instpipe1)
    4'b0011: // mul
      begin
        int_src1datapipe1 =64'h00000000ffffffff
          & postw2re_datapipe3;
        int_src2datapipe1 =64'h00000000ffffffff
          & postw2re_datapipe1;
      end
    4'b1100: // shift left inst.
      begin
        int_src1datapipe1 = postw2re_datapipe3;
        int_src2datapipe1 =64'h000000000000000f
          & postw2re_datapipe1;
      end
    4'b1101: // shift right inst.
      begin
        int_src1datapipe1 = postw2re_datapipe3;
        int_src2datapipe1 =64'h000000000000000f
          & postw2re_datapipe1;
      end
    4'b1110: // barrel shift left inst.
      begin
        int_src1datapipe1 = postw2re_datapipe3;
        int_src2datapipe1 =64'h000000000000000f
          & postw2re_datapipe1;
      end
    4'b1111: // barrel shift right inst.
      begin
        int_src1datapipe1 = postw2re_datapipe3;
        int_src2datapipe1 =64'h000000000000000f
          & postw2re_datapipe1;
      end
    default:
      begin
        int_src1datapipe1 = postw2re_datapipe3;
        int_src2datapipe1 = postw2re_datapipe1;
      end
  endcase
end
// for cross operation register bypass between
// operation1 and operation1 for src1 AND
// between operation3 and operation1 for src2.
```

```verilog
else if ((postw2re_destpipe1 == r2e_src1pipe1) &
  (postw2re_destpipe3 == r2e_src2pipe1))
begin
  case (d2e_instpipe1)
  4'b0011: // mul
    begin
      int_src1datapipe1 =64'h00000000ffffffff
        & postw2re_datapipe1;
      int_src2datapipe1 =64'h00000000ffffffff
        & postw2re_datapipe3;
    end
  4'b1100: // shift left inst.
    begin
      int_src1datapipe1 = postw2re_datapipe1;
      int_src2datapipe1 =64'h000000000000000f
        & postw2re_datapipe3;
    end
  4'b1101: // shift right inst
    begin
      int_src1datapipe1 = postw2re_datapipe1;
      int_src2datapipe1 =64'h000000000000000f
        & postw2re_datapipe3;
    end
  4'b1110: // barrel shift left inst
    begin
      int_src1datapipe1 = postw2re_datapipe1;
      int_src2datapipe1 =64'h000000000000000f
        & postw2re_datapipe3;
    end
  4'b1111: // barrel shift right inst
    begin
      int_src1datapipe1 = postw2re_datapipe1;
      int_src2datapipe1 =64'h000000000000000f
        & postw2re_datapipe3;
    end
  default:
    begin
      int_src1datapipe1 = postw2re_datapipe1;
      int_src2datapipe1 = postw2re_datapipe3;
    end
  endcase
end
// for cross operation register bypass between
// operation1 and operation1 for src1 AND
// between operation1 and operation1 for src2.
else if ((postw2re_destpipe1 == r2e_src1pipe1) &
  (postw2re_destpipe1 == r2e_src2pipe1))
begin
  case (d2e_instpipe1)
  4'b0011: // mul
    begin
      int_src1datapipe1 =64'h00000000ffffffff
        & postw2re_datapipe1;
      int_src2datapipe1 =64'h00000000ffffffff
        & postw2re_datapipe1;
    end
  4'b1100: // shift left inst
    begin
      int_src1datapipe1 = postw2re_datapipe1;
      int_src2datapipe1 =64'h000000000000000f
        & postw2re_datapipe1;
    end
```

```
4'b1101: // shift right inst
  begin
   int_src1datapipe1 = postw2re_datapipe1;
   int_src2datapipe1 =64'h000000000000000f
     & postw2re_datapipe1;
  end
4'b1110: // barrel shift left inst
  begin
   int_src1datapipe1 = postw2re_datapipe1;
   int_src2datapipe1 =64'h000000000000000f
     & postw2re_datapipe1;
  end
4'b1111:  // barrel shift right inst
  begin
   int_src1datapipe1 = postw2re_datapipe1;
   int_src2datapipe1 =64'h000000000000000f
     & postw2re_datapipe1;
  end
default:
  begin
   int_src1datapipe1 = postw2re_datapipe1;
   int_src2datapipe1 = postw2re_datapipe1;
  end
 endcase
 end
// for cross operation register bypass between
// operation2 and operation1 for src1 AND
// between operation2 and operation1 for src2.
else if ((postw2re_destpipe2 == r2e_src2pipe1) &
 (postw2re_destpipe2 == r2e_src1pipe1))
begin
  case (d2e_instpipe1)
    4'b0011: // mul
      begin
        int_src1datapipe1 =64'h00000000ffffffff
          & postw2re_datapipe2;
        int_src2datapipe1 =64'h00000000ffffffff
          & postw2re_datapipe2;
      end
    4'b1100: // shift left inst
      begin
        int_src1datapipe1 = postw2re_datapipe2;
        int_src2datapipe1 =64'h000000000000000f
          & postw2re_datapipe2;
      end
    4'b1101: // shift right inst
      begin
        int_src1datapipe1 = postw2re_datapipe2;
        int_src2datapipe1 =64'h000000000000000f
          & postw2re_datapipe2;
      end
    4'b1110:  // barrel shift left inst
      begin
        int_src1datapipe1 = postw2re_datapipe2;
        int_src2datapipe1 =64'h000000000000000f
          & postw2re_datapipe2;
      end
    4'b1111:  // barrel shift right inst
      begin
        int_src1datapipe1 = postw2re_datapipe2;
        int_src2datapipe1 =64'h000000000000000f
          & postw2re_datapipe2;
      end
```

```
         default:
           begin
             int_src1datapipe1 = postw2re_datapipe2;
             int_src2datapipe1 = postw2re_datapipe2;
           end
       endcase
     end
// for cross operation register bypass between
// operation3 and operation1 for src1 AND
// between operation3 and operation1 for src2.
else if ((postw2re_destpipe3 == r2e_src1pipe1) &
  (postw2re_destpipe3 == r2e_src2pipe1))
  begin
    case (d2e_instpipe1)
      4'b0011: // mul
          begin
            int_src1datapipe1 =64'h00000000ffffffff
              & postw2re_datapipe3;
            int_src2datapipe1 =64'h00000000ffffffff
              & postw2re_datapipe3;
          end
      4'b1100: // shift left inst
          begin
            int_src1datapipe1 = postw2re_datapipe3;
            int_src2datapipe1 =64'h000000000000000f
              & postw2re_datapipe3;
          end
      4'b1101: // shift right inst
          begin
            int_src1datapipe1 = postw2re_datapipe3;
            int_src2datapipe1 =64'h000000000000000f
              & postw2re_datapipe3;
          end
      4'b1110: // barrel shift left inst
          begin
            int_src1datapipe1 = postw2re_datapipe3;
            int_src2datapipe1 =64'h000000000000000f
              & postw2re_datapipe3;
          end
      4'b1111: // barrel shift right inst
          begin
            int_src1datapipe1 = postw2re_datapipe3;
            int_src2datapipe1 =64'h000000000000000f
              & postw2re_datapipe3;
          end
      default:
          begin
            int_src1datapipe1 = postw2re_datapipe3;
            int_src2datapipe1 = postw2re_datapipe3;
          end
    endcase
  end
// for cross operation register bypass between
// operation1 and operation1 for src1 AND
// between operation2 and operation1 for src2.
else if ((postw2re_destpipe2 == r2e_src2pipe1) &
(postw2re_destpipe1 == r2e_src1pipe1))
  begin
    case (d2e_instpipe1)
      4'b0011: // mul
          begin
            int_src1datapipe1 =64'h00000000ffffffff
              & postw2re_datapipe1;
            int_src2datapipe1 =64'h00000000ffffffff
              & postw2re_datapipe2;
          end
```

```verilog
4'b1100: // shift left inst
   begin
     int_src1datapipe1 = postw2re_datapipe1;
     int_src2datapipe1 =64'h000000000000000f
        & postw2re_datapipe2;
   end
4'b1101:  // shift right inst
   begin
     int_src1datapipe1 = postw2re_datapipe1;
     int_src2datapipe1 =64'h000000000000000f
        & postw2re_datapipe2;
   end
4'b1110:  // barrel shift left inst
   begin
     int_src1datapipe1 = postw2re_datapipe1;
     int_src2datapipe1 =64'h000000000000000f
        & postw2re_datapipe2;
   end
4'b1111:  // barrel shift right inst
   begin
     int_src1datapipe1 = postw2re_datapipe1;
     int_src2datapipe1 =64'h000000000000000f
        & postw2re_datapipe2;
   end
default:
   begin
     int_src1datapipe1 = postw2re_datapipe1;
     int_src2datapipe1 = postw2re_datapipe2;
   end
 endcase
end
// for cross operation register bypass between
// operation2 and operation1 for src1 AND
// between operation1 and operation1 for src2.
else if ((postw2re_destpipe1 == r2e_src2pipe1) &
(postw2re_destpipe2 == r2e_src1pipe1))
begin
  case (d2e_instpipe1)
    4'b0011: // mul
       begin
         int_src1datapipe1 =64'h00000000ffffffff
           & postw2re_datapipe2;
         int_src2datapipe1 =64'h00000000ffffffff
           & postw2re_datapipe1;
       end
    4'b1100:  // shift left inst
       begin
         int_src1datapipe1 = postw2re_datapipe2;
         int_src2datapipe1 =64'h000000000000000f
           & postw2re_datapipe1;
       end
    4'b1101:  // shift right inst
       begin
         int_src1datapipe1 = postw2re_datapipe2;
         int_src2datapipe1 =64'h000000000000000f
           & postw2re_datapipe1;
       end
    4'b1110:  // barrel shift left inst
       begin
         int_src1datapipe1 = postw2re_datapipe2;
         int_src2datapipe1 =64'h000000000000000f
           & postw2re_datapipe1;
       end
```

```
       4'b1111:  // barrel shift right inst
         begin
           int_src1datapipe1 = postw2re_datapipe2;
           int_src2datapipe1 =64'h000000000000000f
             & postw2re_datapipe1;
         end
     default:
         begin
           int_src1datapipe1 = postw2re_datapipe2;
           int_src2datapipe1 = postw2re_datapipe1;
         end
   endcase
 end
// for register bypass between operation1 and
// operation1 for src2
else if ((postw2re_destpipe1 == r2e_src2pipe1)
  & ~comp_postw2re_dest)
   begin
     int_src1datapipe1 = r2e_src1datapipe1;
   case (d2e_instpipe1)
     4'b0011: // mul
       int_src2datapipe1 =64'h00000000ffffffff
         & postw2re_datapipe1;
     4'b1100: // shift left inst
       int_src2datapipe1 =64'h000000000000000f
         & postw2re_datapipe1;
     4'b1101: // shift right inst
       int_src2datapipe1 =64'h000000000000000f
         & postw2re_datapipe1;
    4'b1110: // barrel shift left inst
       int_src2datapipe1 =64'h000000000000000f
         & postw2re_datapipe1;
     4'b1111: // barrel shift right inst
       int_src2datapipe1 =64'h000000000000000f
         & postw2re_datapipe1;
     default:
       int_src2datapipe1 = postw2re_datapipe1;
     endcase
 end
// for register bypass between operation1 and
// operation1 for src1
else if ((postw2re_destpipe1 == r2e_src1pipe1)
  & ~comp_postw2re_dest)
 begin
   int_src2datapipe1 = r2e_src2datapipe1;
     case (d2e_instpipe1)
       4'b0011: // mul
         int_src1datapipe1 =64'h00000000ffffffff
           & postw2re_datapipe1;
       4'b1100: // shift left inst
         int_src1datapipe1 = postw2re_datapipe1;
       4'b1101: // shift right inst
         int_src1datapipe1 = postw2re_datapipe1;
       4'b1110: // barrel shift left inst
         int_src1datapipe1 = postw2re_datapipe1;
       4'b1111: // barrel shift right inst
         int_src1datapipe1 = postw2re_datapipe1;
     default:
       int_src1datapipe1 = postw2re_datapipe1;
     endcase
 end
// for cross operation register bypass between
// operation2 and operation1 for src1
else if ((postw2re_destpipe2 == r2e_src1pipe1)
  & ~comp_postw2re_dest)
```

```verilog
begin
  int_src2datapipe1 = r2e_src2datapipe1;
  case (d2e_instpipe1)
    4'b0011: // mul
      int_src1datapipe1 =64'h00000000ffffffff
        & postw2re_datapipe2;
    4'b1100: // shift left inst
      int_src1datapipe1 = postw2re_datapipe2;
    4'b1101: // shift right inst
      int_src1datapipe1 = postw2re_datapipe2;
    4'b1110: // barrel shift left inst
      int_src1datapipe1 = postw2re_datapipe2;
    4'b1111: // barrel shift right inst
      int_src1datapipe1 = postw2re_datapipe2;
    default:
      int_src1datapipe1 = postw2re_datapipe2;
  endcase
end
// for cross operation register bypass between
// operation2 and operation1 for src2
else if ((postw2re_destpipe2 == r2e_src2pipe1)
  & ~comp_postw2re_dest)
begin
 int_src1datapipe1 = r2e_src1datapipe1;
  case (d2e_instpipe1)
    4'b0011: // mul
      int_src2datapipe1 =64'h00000000ffffffff
        & postw2re_datapipe2;
    4'b1100: // shift left inst
      int_src2datapipe1 =64'h000000000000000f
        & postw2re_datapipe2;
    4'b1101: // shift right inst
      int_src2datapipe1 =64'h000000000000000f
        & postw2re_datapipe2;
    4'b1110: // barrel shift left inst
      int_src2datapipe1 =64'h000000000000000f
        & postw2re_datapipe2;
    4'b1111: // barrel shift right inst
      int_src2datapipe1 =64'h000000000000000f
        & postw2re_datapipe2;
    default:
     int_src2datapipe1 = postw2re_datapipe2;
  endcase
end
// for cross operation register bypass between
// operation3 and operation1 for src1
else if ((postw2re_destpipe3 == r2e_src1pipe1)
  & ~comp_postw2re_dest)
begin
 int_src2datapipe1 = r2e_src2datapipe1;
  case (d2e_instpipe1)
    4'b0011: // mul
      int_src1datapipe1 =64'h00000000ffffffff
        & postw2re_datapipe3;
    4'b1100: // shift left inst
      int_src1datapipe1 = postw2re_datapipe3;
    4'b1101: // shift right inst
      int_src1datapipe1 = postw2re_datapipe3;
    4'b1110: // barrel shift left inst
      int_src1datapipe1 = postw2re_datapipe3;
    4'b1111: // barrel shift right inst
      int_src1datapipe1 = postw2re_datapipe3;
```

```
          default:
           int_src1datapipe1 = postw2re_datapipe3;
          endcase
        end
      // for cross operation register bypass between
      // operation3 and operation1 for src2
      else if ((postw2re_destpipe3 == r2e_src2pipe1)
        & ~comp_postw2re_dest)
        begin
         int_src1datapipe1 = r2e_src1datapipe1;
         case (d2e_instpipe1)
           4'b0011: // mul
             int_src2datapipe1 =64'h00000000ffffffff
               & postw2re_datapipe3;
           4'b1100: // shift left inst
             int_src2datapipe1 =64'h000000000000000f
               & postw2re_datapipe3;
           4'b1101: // shift right inst
             int_src2datapipe1 =64'h000000000000000f
               & postw2re_datapipe3;
           4'b1110: // barrel shift left inst
             int_src2datapipe1 =64'h000000000000000f
               & postw2re_datapipe3;
           4'b1111: // barrel shift right inst
             int_src2datapipe1 =64'h000000000000000f
               & postw2re_datapipe3;
           default:
             int_src2datapipe1 = postw2re_datapipe3;
           endcase
        end
      else
        begin
         int_src1datapipe1 = r2e_src1datapipe1;
         int_src2datapipe1 = r2e_src2datapipe1;
        end
    end
end
// for register bypass on operation2
always @ (d2e_instpipe2 or postw2re_destpipe2 or r2e_src1pipe2 or
r2e_src2pipe2 or r2e_src1datapipe2 or r2e_src2datapipe2 or
postw2re_datapipe2 or w2re_destpipe2 or w2re_datapipe2 or
postw2re_datapipe1 or postw2re_destpipe1 or e2w_wrpipe2 or
postw2re_destpipe3 or postw2re_datapipe3) or comp_w2re_dest or
comp_postw2re_dest
begin
  if ((d2e_instpipe2 == load) | (d2e_instpipe2 == nop))
    begin
     int_src1datapipe2 = r2e_src1datapipe2;
     int_src2datapipe2 = r2e_src2datapipe2;
    end
  /* else if (e2w_wrpipe2) // for debug only
    begin
     if (postw2re_destpipe2 == r2e_src1pipe2)
       begin
        int_src1datapipe2 = postw2re_datapipe2;
        int_src2datapipe2 = r2e_src2datapipe2;
       end
     else if (postw2re_destpipe2 == r2e_src2pipe2)
       begin
        int_src1datapipe2 = r2e_src1datapipe2;
        int_src2datapipe2 = postw2re_datapipe2;
       end
     else
```

```
      begin
        int_src1datapipe2 = r2e_src1datapipe2;
        int_src2datapipe2 = r2e_src2datapipe2;
      end
  end */
else
  begin
    if ((w2re_destpipe2 == r2e_src1pipe2)
      & ~comp_w2re_dest)
      begin
        int_src1datapipe2 = w2re_datapipe2;
        int_src2datapipe2 = r2e_src2datapipe2;
      end
    else if ((w2re_destpipe2 == r2e_src2pipe2)
      & ~((w2re_destpipe2== reg0)&(r2e_src2pipe2==reg0)
      &(d2e_instpipe2==read)) &~comp_w2re_dest)
      begin
        int_src1datapipe2 = r2e_src1datapipe2;
        int_src2datapipe2 = w2re_datapipe2;
      end
    // for cross operation register bypass between
    // operation2 and operation2 for src2 AND
    // between operation1 and operation2 for src1.
    else if ((postw2re_destpipe2 == r2e_src2pipe2)
      & (postw2re_destpipe1 == r2e_src1pipe2))
      begin
        case (d2e_instpipe2)
          4'b0011: // mul
            begin
              int_src1datapipe2 =64'h00000000ffffffff
                & postw2re_datapipe1;
              int_src2datapipe2 =64'h00000000ffffffff
                & postw2re_datapipe2;
            end
          4'b1100: // shift left inst
            begin
              int_src1datapipe2 = postw2re_datapipe1;
              int_src2datapipe2 =64'h000000000000000f
                & postw2re_datapipe2;
            end
          4'b1101:  // shift right inst
            begin
              int_src1datapipe2 = postw2re_datapipe1;
              int_src2datapipe2 =64'h000000000000000f
                & postw2re_datapipe2;
            end
          4'b1110:  // barrel shift left inst
            begin
              int_src1datapipe2 = postw2re_datapipe1;
              int_src2datapipe2 =64'h000000000000000f
                & postw2re_datapipe2;
            end
          4'b1111:  // barrel shift right inst
            begin
              int_src1datapipe2 = postw2re_datapipe1;
              int_src2datapipe2 =64'h000000000000000f
                & postw2re_datapipe2;
            end
        default:
        begin
          int_src1datapipe2 = postw2re_datapipe1;
          int_src2datapipe2 = postw2re_datapipe2;
        end
      endcase
    end
```

> This portion of the code is for intrapipe bypass for w2re_destpipe. Extend the "else if (w2re_destpipe" portion of this code for interpipe bypass, similar to the code for the bypass for postw2re_dest pipe below.

```
// for cross operation register bypass between
// operation2 and operation2 for src2 AND
// between operation3 and operation2 for src1.
else if ((postw2re_destpipe2 == r2e_src2pipe2) &
(postw2re_destpipe3 == r2e_src1pipe2))
 begin
   case (d2e_instpipe2)
     4'b0011: // mul
       begin
         int_src1datapipe2 =64'h00000000ffffffff
           & postw2re_datapipe3;
         int_src2datapipe2 =64'h00000000ffffffff
           & postw2re_datapipe2;
       end
     4'b1100:  // shift left inst
       begin
         int_src1datapipe2 = postw2re_datapipe3;
         int_src2datapipe2 =64'h000000000000000f
           & postw2re_datapipe2;
       end
     4'b1101:  // shift right inst
       begin
         int_src1datapipe2 = postw2re_datapipe3;
         int_src2datapipe2 =64'h000000000000000f
           & postw2re_datapipe2;
       end
     4'b1110:  // barrel shift left inst
       begin
         int_src1datapipe2 = postw2re_datapipe3;
         int_src2datapipe2 =64'h000000000000000f
           & postw2re_datapipe2;
       end
     4'b1111:  // barrel shift right inst
       begin
         int_src1datapipe2 = postw2re_datapipe3;
         int_src2datapipe2 =64'h000000000000000f
           & postw2re_datapipe2;
       end
   default:
   begin
     int_src1datapipe2 = postw2re_datapipe3;
     int_src2datapipe2 = postw2re_datapipe2;
   end
   endcase
 end
// for cross operation register bypass between
// operation2 and operation2 for src1 AND
// between operation1 and operation2 for src2.
else if ((postw2re_destpipe2 == r2e_src1pipe2) &
(postw2re_destpipe1 == r2e_src2pipe2))
 begin
   case (d2e_instpipe2)
     4'b0011: // mul
       begin
         int_src1datapipe2 =64'h00000000ffffffff
           & postw2re_datapipe2;
         int_src2datapipe2 =64'h00000000ffffffff
           & postw2re_datapipe1;
       end
```

```
      4'b1100:  // shift left inst
        begin
          int_src1datapipe2 = postw2re_datapipe2;
          int_src2datapipe2 =64'h000000000000000f
            & postw2re_datapipe1;
        end
      4'b1101:  // shift right inst
        begin
          int_src1datapipe2 = postw2re_datapipe2;
          int_src2datapipe2 =64'h000000000000000f
            & postw2re_datapipe1;
        end
      4'b1110:  // barrel shift left inst
        begin
          int_src1datapipe2 = postw2re_datapipe2;
          int_src2datapipe2 =64'h000000000000000f
            & postw2re_datapipe1;
        end
      4'b1111:  // barrel shift right inst
        begin
          int_src1datapipe2 = postw2re_datapipe2;
          int_src2datapipe2 =64'h000000000000000f
            & postw2re_datapipe1;
        end
    default:
      begin
        int_src1datapipe2 = postw2re_datapipe2;
        int_src2datapipe2 = postw2re_datapipe1;
      end
    endcase
  end
// for cross operation register bypass between
// operation2 and operation2 for src1 AND
// between operation3 and operation2 for src2.
else if ((postw2re_destpipe2 == r2e_src1pipe2) &
(postw2re_destpipe3 == r2e_src2pipe2))
 begin
   case (d2e_instpipe2)
     4'b0011: // mul
       begin
         int_src1datapipe2 =64'h00000000ffffffff
           & postw2re_datapipe2;
         int_src2datapipe2 =64'h00000000ffffffff
           & postw2re_datapipe3;
       end
     4'b1100:  // shift left inst
       begin
         int_src1datapipe2 = postw2re_datapipe2;
         int_src2datapipe2 =64'h000000000000000f
           & postw2re_datapipe3;
       end
     4'b1101:  // shift right inst
       begin
         int_src1datapipe2 = postw2re_datapipe2;
         int_src2datapipe2 =64'h000000000000000f
           & postw2re_datapipe3;
       end
     4'b1110:  // barrel shift left inst
       begin
         int_src1datapipe2 = postw2re_datapipe2;
         int_src2datapipe2 =64'h000000000000000f
           & postw2re_datapipe3;
     end
```

```
  4'b1111:  // barrel shift right inst
    begin
      int_src1datapipe2 = postw2re_datapipe2;
      int_src2datapipe2 =64'h000000000000000f
        & postw2re_datapipe3;
    end
  default:
    begin
      int_src1datapipe2 = postw2re_datapipe2;
      int_src2datapipe2 = postw2re_datapipe3;
    end
 endcase
end
// for cross operation register bypass between
// operation2 and operation2 for src1 AND
// between operation2 and operation2 for src2.
else if ((postw2re_destpipe2 == r2e_src1pipe2) &
(postw2re_destpipe2 == r2e_src2pipe2))
 begin
   case (d2e_instpipe2)
     4'b0011:  // mul
       begin
         int_src1datapipe2 =64'h00000000ffffffff
           & postw2re_datapipe2;
         int_src2datapipe2 =64'h00000000ffffffff
           & postw2re_datapipe2;
       end
     4'b1100:  // shift left inst
       begin
         int_src1datapipe2 = postw2re_datapipe2;
         int_src2datapipe2 =64'h000000000000000f
           & postw2re_datapipe2;
       end
     4'b1101:  // shift right inst
       begin
         int_src1datapipe2 = postw2re_datapipe2;
         int_src2datapipe2 =64'h000000000000000f
           & postw2re_datapipe2;
       end
     4'b1110:  // barrel shift left inst
       begin
         int_src1datapipe2 = postw2re_datapipe2;
         int_src2datapipe2 =64'h000000000000000f
           & postw2re_datapipe2;
       end
     4'b1111:  // barrel shift right inst
       begin
         int_src1datapipe2 = postw2re_datapipe2;
         int_src2datapipe2 =64'h000000000000000f
           & postw2re_datapipe2;
       end
     default:
       begin
         int_src1datapipe2 = postw2re_datapipe2;
         int_src2datapipe2 = postw2re_datapipe2;
       end
   endcase
 end
```

```verilog
// for cross operation register bypass between
// operation1 and operation2 for src1 AND
// between operation1 and operation2 for src2.
else if ((postw2re_destpipe1 == r2e_src2pipe2) &
(postw2re_destpipe1 == r2e_src1pipe2))
 begin
   case (d2e_instpipe2)
     4'b0011: // mul
       begin
         int_src1datapipe2 =64'h00000000ffffffff
           & postw2re_datapipe1;
         int_src2datapipe2 =64'h00000000ffffffff
           & postw2re_datapipe1;
       end
     4'b1100: // shift left inst
       begin
         int_src1datapipe2 = postw2re_datapipe1;
         int_src2datapipe2 =64'h000000000000000f
           & postw2re_datapipe1;
       end
     4'b1101: // shift right inst
       begin
         int_src1datapipe2 = postw2re_datapipe1;
         int_src2datapipe2 =64'h000000000000000f
           & postw2re_datapipe1;
       end
     4'b1110: // barrel shift left inst
       begin
         int_src1datapipe2 = postw2re_datapipe1;
         int_src2datapipe2 =64'h000000000000000f
           & postw2re_datapipe1;
       end
     4'b1111: // barrel shift right inst
       begin
         int_src1datapipe2 = postw2re_datapipe1;
         int_src2datapipe2 =64'h000000000000000f
           & postw2re_datapipe1;
       end
    default:
       begin
         int_src1datapipe2 = postw2re_datapipe1;
         int_src2datapipe2 = postw2re_datapipe1;
       end
  endcase
 end
// for cross operation register bypass between
// operation3 and operation2 for src1 AND
// between operation3 and operation2 for src2.
else if ((postw2re_destpipe3 == r2e_src2pipe2) &
(postw2re_destpipe3 == r2e_src1pipe2))
 begin
   case (d2e_instpipe2)
     4'b0011: // mul
       begin
         int_src1datapipe2 =64'h00000000ffffffff
           & postw2re_datapipe3;
         int_src2datapipe2 =64'h00000000ffffffff
           & postw2re_datapipe3;
       end
```

```
   4'b1100:  // shift left inst
     begin
       int_src1datapipe2 = postw2re_datapipe3;
       int_src2datapipe2 =64'h000000000000000f
         & postw2re_datapipe3;
     end
   4'b1101:  // shift right inst
     begin
       int_src1datapipe2 = postw2re_datapipe3;
       int_src2datapipe2 =64'h000000000000000f
         & postw2re_datapipe3;
     end
   4'b1110:  // barrel shift left inst
     begin
       int_src1datapipe2 = postw2re_datapipe3;
       int_src2datapipe2 =64'h000000000000000f
         & postw2re_datapipe3;
     end
   4'b1111:  // barrel shift right inst
     begin
       int_src1datapipe2 = postw2re_datapipe3;
       int_src2datapipe2 =64'h000000000000000f
         & postw2re_datapipe3;
     end
  default:
     begin
       int_src1datapipe2 = postw2re_datapipe3;
       int_src2datapipe2 = postw2re_datapipe3;
     end
 endcase
end
// for cross operation register bypass between
// operation3 and operation2 for src2 AND
// between operation1 and operation2 for src1.
else if ((postw2re_destpipe3 == r2e_src2pipe2) &
(postw2re_destpipe1 == r2e_src1pipe2))
 begin
   case (d2e_instpipe2)
     4'b0011: // mul
       begin
         int_src1datapipe2 =64'h00000000ffffffff
           & postw2re_datapipe1;
         int_src2datapipe2 =64'h00000000ffffffff
           & postw2re_datapipe3;
       end
     4'b1100: // shift left inst
       begin
         int_src1datapipe2 = postw2re_datapipe1;
         int_src2datapipe2 =64'h000000000000000f
           & postw2re_datapipe3;
       end
     4'b1101:  // shift right inst
       begin
         int_src1datapipe2 = postw2re_datapipe1;
         int_src2datapipe2 =64'h000000000000000f
           & postw2re_datapipe3;
       end
     4'b1110:  // barrel shift left inst
       begin
         int_src1datapipe2 = postw2re_datapipe1;
         int_src2datapipe2 =64'h000000000000000f
           & postw2re_datapipe3;
       end
```

```
        4'b1111:  // barrel shift right inst
          begin
            int_src1datapipe2 = postw2re_datapipe1;
            int_src2datapipe2 =64'h000000000000000f
              & postw2re_datapipe3;
          end
      default:
          begin
            int_src1datapipe2 = postw2re_datapipe1;
            int_src2datapipe2 = postw2re_datapipe3;
          end
    endcase
  end
// for cross operation register bypass between
// operation1 and operation2 for src2 AND
// between operation3 and operation2 for src1.
else if ((postw2re_destpipe1 == r2e_src2pipe2) &
(postw2re_destpipe3 == r2e_src1pipe2))
  begin
    case (d2e_instpipe2)
      4'b0011: // mul
          begin
            int_src1datapipe2 =64'h00000000ffffffff
              & postw2re_datapipe3;
            int_src2datapipe2 =64'h00000000ffffffff
              & postw2re_datapipe1;
          end
      4'b1100:  // shift left inst
          begin
            int_src1datapipe2 = postw2re_datapipe3;
            int_src2datapipe2 =64'h000000000000000f
              & postw2re_datapipe1;
          end
      4'b1101:  // shift right inst
          begin
            int_src1datapipe2 = postw2re_datapipe3;
            int_src2datapipe2 =64'h000000000000000f
              & postw2re_datapipe1;
          end
      4'b1110:  // barrel shift left inst
          begin
            int_src1datapipe2 = postw2re_datapipe3;
            int_src2datapipe2 =64'h000000000000000f
              & postw2re_datapipe1;
          end
      4'b1111:  // barrel shift right inst
          begin
            int_src1datapipe2 = postw2re_datapipe3;
            int_src2datapipe2 =64'h000000000000000f
              & postw2re_datapipe1;
          end
      default:
          begin
            int_src1datapipe2 = postw2re_datapipe3;
            int_src2datapipe2 = postw2re_datapipe1;
          end
    endcase
  end
// for cross operation register bypass between
// operation1 and operation2 for src2
else if ((postw2re_destpipe1 == r2e_src2pipe2)
  & ~comp_postw2re_dest)
```

```
begin
  int_src1datapipe2 = r2e_src1datapipe2;
    case (d2e_instpipe2)
      4'b0011: // mul
        int_src2datapipe2 = 64'h00000000ffffffff &
        postw2re_datapipe1;
      4'b1100: // shift left inst
        int_src2datapipe2 = 64'h000000000000000f &
        postw2re_datapipe1;
      4'b1101: // shift right inst
        int_src2datapipe2 = 64'h000000000000000f &
        postw2re_datapipe1;
      4'b1110: // barrel shift left inst
        int_src2datapipe2 = 64'h000000000000000f &
        postw2re_datapipe1;
      4'b1111: // barrel shift right inst
        int_src2datapipe2 = 64'h000000000000000f &
        postw2re_datapipe1;
      default:
        int_src2datapipe2 = postw2re_datapipe1;
    endcase
 end
// for cross operation register bypass between
// operation1 and operation2 for src1
else if ((postw2re_destpipe1 == r2e_src1pipe2)
 & ~comp_postw2re_dest)
 begin
 int_src2datapipe2 = r2e_src2datapipe2;
    case (d2e_instpipe2)
      4'b0011: // mul
        int_src1datapipe2 = 64'h00000000ffffffff &
        postw2re_datapipe1;
      4'b1100: // shift left inst
        int_src1datapipe2 = postw2re_datapipe1;
      4'b1101: // shift right inst
        int_src1datapipe2 = postw2re_datapipe1;
      4'b1110: // barrel shift left inst
        int_src1datapipe2 = postw2re_datapipe1;
      4'b1111: // barrel shift right inst
        int_src1datapipe2 = postw2re_datapipe1;
      default:
        int_src1datapipe2 = postw2re_datapipe1;
    endcase
 end
// for register bypass between operation2 and
// operation2 for src2
else if ((postw2re_destpipe2 == r2e_src2pipe2)
 & ~comp_postw2re_dest)
 begin
 int_src1datapipe2 = r2e_src1datapipe2;
    case (d2e_instpipe2)
      4'b0011: // mul
        int_src2datapipe2 = 64'h00000000ffffffff &
        postw2re_datapipe2;
      4'b1100: // shift left inst
        int_src2datapipe2 = 64'h000000000000000f &
        postw2re_datapipe2;
      4'b1101: // shift right inst
        int_src2datapipe2 = 64'h000000000000000f &
        postw2re_datapipe2;
      4'b1110: // barrel shift left inst
        int_src2datapipe2 = 64'h000000000000000f &
        postw2re_datapipe2;
```

```
4'b1111:  // barrel shift right inst
  int_src2datapipe2 = 64'h000000000000000f &
    postw2re_datapipe2;
default:
  int_src2datapipe2 = postw2re_datapipe2;
endcase
end
// for register bypass between operation2 and
// operation2 for src1
else if ((postw2re_destpipe2 == r2e_src1pipe2)
 & ~comp_postw2re_dest)
begin
  int_src2datapipe2 = r2e_src2datapipe2;
  case (d2e_instpipe2)
    4'b0011: // mul
      int_src1datapipe2 = 64'h00000000ffffffff &
        postw2re_datapipe2;
    4'b1100:  // shift left inst
      int_src1datapipe2 = postw2re_datapipe2;
    4'b1101:  // shift right inst
      int_src1datapipe2 = postw2re_datapipe2;
    4'b1110:  // barrel shift left inst
      int_src1datapipe2 = postw2re_datapipe2;
    4'b1111:  // barrel shift right inst
      int_src1datapipe2 = postw2re_datapipe2;
    default:
      int_src1datapipe2 = postw2re_datapipe2;
  endcase
end
// for cross operation register bypass between
// operation3 and operation2 for src1
else if ((postw2re_destpipe3 == r2e_src1pipe2)
 & ~comp_postw2re_dest)
begin
 int_src2datapipe2 = r2e_src2datapipe2;
  case (d2e_instpipe2)
    4'b0011: // mul
      int_src1datapipe2 = 64'h00000000ffffffff &
        postw2re_datapipe3;
    4'b1100:  // shift left inst
      int_src1datapipe2 = postw2re_datapipe3;
    4'b1101: // shift right inst
      int_src1datapipe2 = postw2re_datapipe3;
    4'b1110: // barrel shift left inst
      int_src1datapipe2 = postw2re_datapipe3;
    4'b1111:  // barrel shift right inst
      int_src1datapipe2 = postw2re_datapipe3;
    default:
      int_src1datapipe2 = postw2re_datapipe3;
  endcase
end
// for cross operation register bypass between
// operation3 and operation2 for src2
else if ((postw2re_destpipe3 == r2e_src2pipe2)
 & ~comp_postw2re_dest)
begin
  int_src1datapipe2 = r2e_src1datapipe2;
    case (d2e_instpipe2)
      4'b0011: // mul
        int_src2datapipe2 = 64'h00000000ffffffff &
          postw2re_datapipe3;
      4'b1100:  // shift left inst
        int_src2datapipe2 = 64'h000000000000000f &
          postw2re_datapipe3;
```

```
              4'b1101:  // shift right inst
                int_src2datapipe2 = 64'h000000000000000f &
                  postw2re_datapipe3;
              4'b1110:  // barrel shift left inst
                int_src2datapipe2 = 64'h000000000000000f &
                  postw2re_datapipe3;
              4'b1111:  // barrel shift right inst
                int_src2datapipe2 = 64'h000000000000000f &
                  postw2re_datapipe3;
              default:
                int_src2datapipe2 = postw2re_datapipe3;
            endcase
        end
      else
        begin
        int_src1datapipe2 = r2e_src1datapipe2;
        int_src2datapipe2 = r2e_src2datapipe2;
        end
    end
end
// for register bypass on operation3
always @ (d2e_instpipe3 or postw2re_destpipe3 or r2e_src1pipe3 or
r2e_src2pipe3 or r2e_src1datapipe3 or r2e_src2datapipe3 or
postw2re_datapipe3 or w2re_destpipe3 or w2re_datapipe3 or
postw2re_datapipe1 or postw2re_destpipe1 or e2w_wrpipe3 or postw2re_
destpipe2 or postw2re_datapipe2 or comp_w2re_dest or comp_postw2re_dest)
begin
  if ((d2e_instpipe3 == load) | (d2e_instpipe3 == nop))
    begin
      int_src1datapipe3 = r2e_src1datapipe3;
      int_src2datapipe3 = r2e_src2datapipe3;
    end
  /* else if (e2w_wrpipe3) // for debug only
    begin
      if (postw2re_destpipe3 == r2e_src1pipe3)
        begin
          int_src1datapipe3 = postw2re_datapipe3;
          int_src2datapipe3 = r2e_src2datapipe3;
        end
      else if (postw2re_destpipe3 == r2e_src2pipe3)
        begin
          int_src1datapipe3 = r2e_src1datapipe3;
          int_src2datapipe3 = postw2re_datapipe3;
        end
      else
        begin
          int_src1datapipe3 = r2e_src1datapipe3;
          int_src2datapipe3 = r2e_src2datapipe3;
        end
    end */
  else
    begin
      if ((w2re_destpipe3 == r2e_src1pipe3)
        & ~comp_w2re_dest)
        begin
          int_src1datapipe3 = w2re_datapipe3;
          int_src2datapipe3 = r2e_src2datapipe3;
        end
      else if ((w2re_destpipe3 == r2e_src2pipe3)
        & ~((w2re_destpipe3 == reg0)
        &(r2e_src2pipe3==reg0) &(d2e_instpipe3==read))
        & ~comp_w2re_dest)
        begin
          int_src1datapipe3 = r2e_src1datapipe3;
          int_src2datapipe3 = w2re_datapipe3;
        end
```

This portion of the code is for intrapipe bypass for w2re_destpipe. Extend the "else if (w2re_destpipe" portion of this code for interpipe bypass, similar to the code for the bypass for postw2re_destpipe below.

```
// for cross operation register bypass between
// operation3 and operation3 for src1 AND
// between operation2 and operation3 for src2.
else if ((postw2re_destpipe3 == r2e_src1pipe3) &
(postw2re_destpipe2 == r2e_src2pipe3))
 begin
   case (d2e_instpipe3)
     4'b0011: // mul
       begin
         int_src1datapipe3 =64'h00000000ffffffff
           & postw2re_datapipe3;
         int_src2datapipe3 =64'h00000000ffffffff
           & postw2re_datapipe2;
       end
     4'b1100: // shift left inst
       begin
         int_src1datapipe3 = postw2re_datapipe3;
         int_src2datapipe3 =64'h000000000000000f
           & postw2re_datapipe2;
       end
     4'b1101: // shift right inst
       begin
         int_src1datapipe3 = postw2re_datapipe3;
         int_src2datapipe3 =64'h000000000000000f
           & postw2re_datapipe2;
       end
     4'b1110: // barrel shift left inst
       begin
         int_src1datapipe3 = postw2re_datapipe3;
         int_src2datapipe3 =64'h000000000000000f
           & postw2re_datapipe2;
       end
     4'b1111: // barrel shift right inst
       begin
         int_src1datapipe3 = postw2re_datapipe3;
         int_src2datapipe3 =64'h000000000000000f
           & postw2re_datapipe2;
       end
   default:
       begin
         int_src1datapipe3 = postw2re_datapipe3;
         int_src2datapipe3 = postw2re_datapipe2;
       end
   endcase
 end
// for cross operation register bypass between
// operation3 and operation3 for src2 AND
// between operation2 and operation3 for src1.
else if ((postw2re_destpipe3 == r2e_src2pipe3) &
(postw2re_destpipe2 == r2e_src1pipe3))
 begin
   case (d2e_instpipe3)
     4'b0011: // mul
       begin
         int_src1datapipe3 =64'h00000000ffffffff
           & postw2re_datapipe2;
         int_src2datapipe3 =64'h00000000ffffffff
           & postw2re_datapipe3;
       end
```

```
      4'b1100:  // shift left inst
        begin
          int_src1datapipe3 = postw2re_datapipe2;
          int_src2datapipe3 =64'h000000000000000f
            & postw2re_datapipe3;
        end
      4'b1101:  // shift right inst
        begin
          int_src1datapipe3 = postw2re_datapipe2;
          int_src2datapipe3 =64'h000000000000000f
            & postw2re_datapipe3;
        end
      4'b1110:  // barrel shift left inst
        begin
          int_src1datapipe3 = postw2re_datapipe2;
          int_src2datapipe3 =64'h000000000000000f
            & postw2re_datapipe3;
        end
      4'b1111:  // barrel shift right inst
        begin
          int_src1datapipe3 = postw2re_datapipe2;
          int_src2datapipe3 =64'h000000000000000f
            & postw2re_datapipe3;
        end
    default:
        begin
          int_src1datapipe3 = postw2re_datapipe2;
          int_src2datapipe3 = postw2re_datapipe3;
        end
    endcase
 end
// for cross operation register bypass between
// operation3 and operation3 for src2 AND
// between operation1 and operation3 for src1.
else if ((postw2re_destpipe3 == r2e_src2pipe3) &
(postw2re_destpipe1 == r2e_src1pipe3))
 begin
   case (d2e_instpipe3)
     4'b0011: // mul
       begin
         int_src1datapipe3 =64'h00000000ffffffff
           & postw2re_datapipe1;
         int_src2datapipe3 =64'h00000000ffffffff
           & postw2re_datapipe3;
       end
     4'b1100:  // shift left inst
       begin
         int_src1datapipe3 = postw2re_datapipe1;
         int_src2datapipe3 =64'h000000000000000f
           & postw2re_datapipe3;
       end
     4'b1101:  // shift right inst
       begin
         int_src1datapipe3 = postw2re_datapipe1;
         int_src2datapipe3 =64'h000000000000000f
           & postw2re_datapipe3;
       end
     4'b1110:  // barrel shift left inst
       begin
         int_src1datapipe3 = postw2re_datapipe1;
         int_src2datapipe3 =64'h000000000000000f
           & þostw2re_datapipe3;
       end
   end
```

```verilog
        4'b1111:  // barrel shift right inst
          begin
            int_src1datapipe3 = postw2re_datapipe1;
            int_src2datapipe3 =64'h000000000000000f
              & postw2re_datapipe3;
          end
      default:
          begin
            int_src1datapipe3 = postw2re_datapipe1;
            int_src2datapipe3 = postw2re_datapipe3;
          end
      endcase
   end
// for cross operation register bypass between
// operation3 and operation3 for src1 AND
// between operation1 and operation3 for src2.
   else if ((postw2re_destpipe3 == r2e_src1pipe3) &
   (postw2re_destpipe1 == r2e_src2pipe3))
    begin
      case (d2e_instpipe3)
        4'b0011: // mul
          begin
            int_src1datapipe3 =64'h00000000ffffffff
              & postw2re_datapipe3;
            int_src2datapipe3 =64'h00000000ffffffff
              & postw2re_datapipe1;
          end
        4'b1100:  // shift left inst
          begin
            int_src1datapipe3 = postw2re_datapipe3;
            int_src2datapipe3 =64'h000000000000000f
              & postw2re_datapipe1;
          end
        4'b1101:  // shift right inst
          begin
            int_src1datapipe3 = postw2re_datapipe3;
            int_src2datapipe3 =64'h000000000000000f
              & postw2re_datapipe1;
          end
        4'b1110:  // barrel shift left inst
          begin
            int_src1datapipe3 = postw2re_datapipe3;
            int_src2datapipe3 =64'h000000000000000f
              & postw2re_datapipe1;
          end
        4'b1111:  // barrel shift right inst
          begin
            int_src1datapipe3 = postw2re_datapipe3;
            int_src2datapipe3 =64'h000000000000000f
              & postw2re_datapipe1;
          end
      default:
          begin
            int_src1datapipe3 = postw2re_datapipe3;
            int_src2datapipe3 = postw2re_datapipe1;
          end
      endcase
   end
```

```verilog
// for cross operation register bypass between
// operation3 and operation3 for src1 AND
// between operation3 and operation3 for src2.
else if ((postw2re_destpipe3 == r2e_src1pipe3) &
(postw2re_destpipe3 == r2e_src2pipe3))
 begin
   case (d2e_instpipe3)
     4'b0011: // mul
       begin
         int_src1datapipe3 =64'h00000000ffffffff
           & postw2re_datapipe3;
         int_src2datapipe3 =64'h00000000ffffffff
           & postw2re_datapipe3;
       end
     4'b1100: // shift left inst
       begin
         int_src1datapipe3 = postw2re_datapipe3;
         int_src2datapipe3 =64'h000000000000000f
           & postw2re_datapipe3;
       end
     4'b1101: // shift right inst
       begin
         int_src1datapipe3 = postw2re_datapipe3;
         int_src2datapipe3 =64'h000000000000000f
           & postw2re_datapipe3;
       end
     4'b1110: // barrel shift left inst
       begin
         int_src1datapipe3 = postw2re_datapipe3;
         int_src2datapipe3 =64'h000000000000000f
           & postw2re_datapipe3;
       end
     4'b1111: // barrel shift right inst
       begin
         int_src1datapipe3 = postw2re_datapipe3;
         int_src2datapipe3 =64'h000000000000000f
           & postw2re_datapipe3;
       end
     default:
       begin
         int_src1datapipe3 = postw2re_datapipe3;
         int_src2datapipe3 = postw2re_datapipe3;
       end
   endcase
 end
// for cross operation register bypass between
// operation1 and operation3 for src1 AND
// between operation1 and operation3 for src2.
else if ((postw2re_destpipe1 == r2e_src1pipe3) &
(postw2re_destpipe1 == r2e_src2pipe3))
 begin
   case (d2e_instpipe3)
     4'b0011: // mul
       begin
         int_src1datapipe3 =64'h00000000ffffffff
           & postw2re_datapipe1;
         int_src2datapipe3 =64'h00000000ffffffff
           & postw2re_datapipe1;
       end
```

```verilog
4'b1100: // shift left inst
  begin
    int_src1datapipe3 = postw2re_datapipe1;
    int_src2datapipe3 =64'h000000000000000f
      & postw2re_datapipe1;
  end
4'b1101:  // shift right inst
  begin
    int_src1datapipe3 = postw2re_datapipe1;
    int_src2datapipe3 =64'h000000000000000f
      & postw2re_datapipe1;
  end
4'b1110:  // barrel shift left inst
  begin
    int_src1datapipe3 = postw2re_datapipe1;
    int_src2datapipe3 =64'h000000000000000f
      & postw2re_datapipe1;
  end
4'b1111:  // barrel shift right inst
  begin
    int_src1datapipe3 = postw2re_datapipe1;
    int_src2datapipe3 =64'h000000000000000f
      & postw2re_datapipe1;
  end
default:
  begin
    int_src1datapipe3 = postw2re_datapipe1;
    int_src2datapipe3 = postw2re_datapipe1;
  end
endcase
end
// for cross operation register bypass between
// operation2 and operation3 for src1 AND
// between operation2 and operation3 for src2.
else if ((postw2re_destpipe2 == r2e_src1pipe3) &
(postw2re_destpipe2 == r2e_src2pipe3))
 begin
  case (d2e_instpipe3)
    4'b0011: // mul
      begin
        int_src1datapipe3 =64'h00000000ffffffff
          & postw2re_datapipe2;
        int_src2datapipe3 =64'h00000000ffffffff
          & postw2re_datapipe2;
      end
    4'b1100:  // shift left inst
      begin
        int_src1datapipe3 = postw2re_datapipe2;
        int_src2datapipe3 =64'h000000000000000f
          & postw2re_datapipe2;
      end
    4'b1101:  // shift right inst
      begin
        int_src1datapipe3 = postw2re_datapipe2;
        int_src2datapipe3 =64'h000000000000000f
          & postw2re_datapipe2;
      end
    4'b1110:  // barrel shift left inst
      begin
        int_src1datapipe3 = postw2re_datapipe2;
        int_src2datapipe3 =64'h000000000000000f
          & postw2re_datapipe2;
      end
```

```
    4'b1111:  // barrel shift right inst
      begin
        int_src1datapipe3 = postw2re_datapipe2;
        int_src2datapipe3 =64'h000000000000000f
          & postw2re_datapipe2;
      end
  default:
      begin
        int_src1datapipe3 = postw2re_datapipe2;
        int_src2datapipe3 = postw2re_datapipe2;
      end
  endcase
end
// for cross operation register bypass between
// operation1 and operation3 for src1 AND
// between operation2 and operation3 for src2.
else if ((postw2re_destpipe1 == r2e_src1pipe3) &
(postw2re_destpipe2 == r2e_src2pipe3))
begin
  case (d2e_instpipe3)
    4'b0011: // mul
      begin
        int_src1datapipe3 =64'h00000000ffffffff
          & postw2re_datapipe1;
        int_src2datapipe3 =64'h00000000ffffffff
          & postw2re_datapipe2;
      end
    4'b1100:  // shift left inst
      begin
        int_src1datapipe3 = postw2re_datapipe1;
        int_src2datapipe3 =64'h000000000000000f
          & postw2re_datapipe2;
      end
    4'b1101:  // shift right inst
      begin
        int_src1datapipe3 = postw2re_datapipe1;
        int_src2datapipe3 =64'h000000000000000f
          & postw2re_datapipe2;
      end
    4'b1110:  // barrel shift left inst
      begin
        int_src1datapipe3 = postw2re_datapipe1;
        int_src2datapipe3 =64'h000000000000000f
          & postw2re_datapipe2;
      end
    4'b1111:  // barrel shift right inst
      begin
        int_src1datapipe3 = postw2re_datapipe1;
        int_src2datapipe3 =64'h000000000000000f
          & postw2re_datapipe2;
      end
  default:
      begin
        int_src1datapipe3 = postw2re_datapipe1;
        int_src2datapipe3 = postw2re_datapipe2;
      end
  endcase
end
```

```verilog
// for cross operation register bypass between
// operation2 and operation3 for src1 AND
// between operation1 and operation3 for src2.
else if ((postw2re_destpipe2 == r2e_src1pipe3) &
(postw2re_destpipe1 == r2e_src2pipe3))
 begin
   case (d2e_instpipe3)
     4'b0011: // mul
       begin
         int_src1datapipe3 =64'h00000000ffffffff
           & postw2re_datapipe2;
         int_src2datapipe3 =64'h00000000ffffffff
           & postw2re_datapipe1;
       end
     4'b1100:  // shift left inst
       begin
         int_src1datapipe3 = postw2re_datapipe2;
         int_src2datapipe3 =64'h000000000000000f
           & postw2re_datapipe1;
       end
     4'b1101:  // shift right inst
       begin
         int_src1datapipe3 = postw2re_datapipe2;
         int_src2datapipe3 =64'h000000000000000f
           & postw2re_datapipe1;
       end
     4'b1110:  // barrel shift left inst
       begin
         int_src1datapipe3 = postw2re_datapipe2;
         int_src2datapipe3 =64'h000000000000000f
           & postw2re_datapipe1;
       end
     4'b1111:  // barrel shift right inst
       begin
         int_src1datapipe3 = postw2re_datapipe2;
         int_src2datapipe3 =64'h000000000000000f
           & postw2re_datapipe1;
       end
   default:
       begin
         int_src1datapipe3 = postw2re_datapipe2;
         int_src2datapipe3 = postw2re_datapipe1;
       end
   endcase
 end
// for cross operation register bypass between
// operation1 and operation3 for src1
else if ((postw2re_destpipe1 == r2e_src1pipe3)
 & ~comp_postw2re_dest)
 begin
  int_src2datapipe3 = r2e_src2datapipe3;
   case (d2e_instpipe3)
     4'b0011: // mul
       int_src1datapipe3 = 64'h00000000ffffffff &
       postw2re_datapipe1;
     4'b1100:  // shift left inst
       int_src1datapipe3 = postw2re_datapipe1;
     4'b1101:  // shift right inst
       int_src1datapipe3 = postw2re_datapipe1;
     4'b1110:  // barrel shift left inst
       int_src1datapipe3 = postw2re_datapipe1;
```

```
      4'b1111:  // barrel shift right inst
        int_src1datapipe3 = postw2re_datapipe1;
     default:
        int_src1datapipe3 = postw2re_datapipe1;
     endcase
    end
// for cross operation register bypass between
// operation1 and operation3 for src2
else if ((postw2re_destpipe1 == r2e_src2pipe3)
 & ~comp_postw2re_dest)
 begin
  int_src1datapipe3 = r2e_src1datapipe3;
    case (d2e_instpipe3)
      4'b0011: // mul
        int_src2datapipe3 = 64'h00000000ffffffff &
        postw2re_datapipe1;
      4'b1100:  // shift left inst
        int_src2datapipe3 = 64'h000000000000000f &
        postw2re_datapipe1;
      4'b1101:  // shift right inst
        int_src2datapipe3 = 64'h000000000000000f &
        postw2re_datapipe1;
      4'b1110:  // barrel shift left inst
        int_src2datapipe3 = 64'h000000000000000f &
        postw2re_datapipe1;
      4'b1111:  // barrel shift right inst
        int_src2datapipe3 = 64'h000000000000000f &
        postw2re_datapipe1;
     default:
        int_src2datapipe3 = postw2re_datapipe1;
     endcase
    end
// for cross operation register bypass between
// operation2 and operation3 for src1
else if ((postw2re_destpipe2 == r2e_src1pipe3)
  & ~comp_postw2re_dest)
 begin
  int_src2datapipe3 = r2e_src2datapipe3;
    case (d2e_instpipe3)
      4'b0011: // mul
        int_src1datapipe3 = 64'h00000000ffffffff &
        postw2re_datapipe2;
      4'b1100:  // shift left inst
        int_src1datapipe3 = postw2re_datapipe2;
      4'b1101:  // shift right inst
        int_src1datapipe3 = postw2re_datapipe2;
      4'b1110:  // barrel shift left inst
        int_src1datapipe3 = postw2re_datapipe2;
      4'b1111:  // barrel shift right inst
        int_src1datapipe3 = postw2re_datapipe2;
     default:
        int_src1datapipe3 = postw2re_datapipe2;
     endcase
    end
// for cross operation register bypass between
// operation2 and operation3 for src2
else if ((postw2re_destpipe2 == r2e_src2pipe3)
  & ~comp_postw2re_dest)
 begin
  int_src1datapipe3 = r2e_src1datapipe3;
    case (d2e_instpipe3)
      4'b0011: // mul
        int_src2datapipe3 = 64'h00000000ffffffff &
        postw2re_datapipe2;
```

```
      4'b1100:  // shift left inst
        int_src2datapipe3 = 64'h000000000000000f &
        postw2re_datapipe2;
      4'b1101:  // shift right inst
        int_src2datapipe3 = 64'h000000000000000f &
        postw2re_datapipe2;
      4'b1110:  // barrel shift left inst
        int_src2datapipe3 = 64'h000000000000000f &
        postw2re_datapipe2;
      4'b1111:  // barrel shift right inst
        int_src2datapipe3 = 64'h000000000000000f &
        postw2re_datapipe2;
      default:
        int_src2datapipe3 = postw2re_datapipe2;
      endcase
 end
// for register bypass between operation3 and
// operation3 for src1
else if ((postw2re_destpipe3 == r2e_src1pipe3)
 & ~comp_postw2re_dest)
 begin
  int_src2datapipe3 = r2e_src2datapipe3;
    case (d2e_instpipe3)
     4'b0011: // mul
        int_src1datapipe3 = 64'h00000000ffffffff &
        postw2re_datapipe3;
     4'b1100:  // shift left inst
        int_src1datapipe3 = postw2re_datapipe3;
     4'b1101:  // shift right inst
        int_src1datapipe3 = postw2re_datapipe3;
     4'b1110:  // barrel shift left inst
        int_src1datapipe3 = postw2re_datapipe3;
     4'b1111:  // barrel shift right inst
        int_src1datapipe3 = postw2re_datapipe3;
     default:
        int_src1datapipe3 = postw2re_datapipe3;
     endcase
 end
// for register bypass between operation3 and
// operation3 for src2
else if ((postw2re_destpipe3 == r2e_src2pipe3)
 & ~comp_postw2re_dest)
 begin
  int_src1datapipe3 = r2e_src1datapipe3;
    case (d2e_instpipe3)
     4'b0011: // mul
        int_src2datapipe3 = 64'h00000000ffffffff &
        postw2re_datapipe3;
     4'b1100:  // shift left inst
        int_src2datapipe3 = 64'h000000000000000f &
        postw2re_datapipe3;
     4'b1101:  // shift right inst
        int_src2datapipe3 = 64'h000000000000000f &
        postw2re_datapipe3;
     4'b1110:  // barrel shift left inst
        int_src2datapipe3 = 64'h000000000000000f &
        postw2re_datapipe3;
     4'b1111:  // barrel shift right inst
        int_src2datapipe3 = 64'h000000000000000f &
        postw2re_datapipe3;
     default:
        int_src2datapipe3 = postw2re_datapipe3;
     endcase
   end
```

```
        else
          begin
            int_src1datapipe3 = r2e_src1datapipe3;
            int_src2datapipe3 = r2e_src2datapipe3;
          end
      end
end
always @ (posedge clock or posedge reset)
begin
  if (reset)
  begin
    e2w_destpipe1 <= reg0;
    e2w_destpipe2 <= reg0;
    e2w_destpipe3 <= reg0;
    e2w_datapipe1 <= 0;
    e2w_datapipe2 <= 0;
    e2w_datapipe3 <= 0;
    e2w_wrpipe1 <= 0;
    e2w_wrpipe2 <= 0;
    e2w_wrpipe3 <= 0;
    e2w_readpipe1 <= 0;
    e2w_readpipe2 <= 0;
    e2w_readpipe3 <= 0;
    preflush <= 0;
    jump <= 0;
  end
  else // positive edge of clock detected
  begin
      // execute for operation 1 pipe1
      case (d2e_instpipe1)
        nop:
          begin
            // in noop, all default to zero
            e2w_destpipe1 <= reg0;
            e2w_datapipe1 <= 0;
            e2w_wrpipe1 <= 0;
            e2w_readpipe1 <= 0;
          end
        add:
          begin
            // src1 + src2 -> dest
            e2w_destpipe1 <= d2e_destpipe1;
            e2w_datapipe1 <= int_src1datapipe1
              + int_src2datapipe1;
            e2w_wrpipe1 <= 1;
            e2w_readpipe1 <= 0;
          end
        sub:
          begin
            // src1 - src2 -> dest
            e2w_destpipe1 <= d2e_destpipe1;
            e2w_datapipe1 <= int_src1datapipe1
              - int_src2datapipe1;
            e2w_wrpipe1 <= 1;
            e2w_readpipe1 <= 0;
          end
        mul:
          begin
            // src1 x src2 -> dest
            // only 32 bits considered
            e2w_destpipe1 <= d2e_destpipe1;
            e2w_datapipe1 <= int_src1datapipe1
              * int_src2datapipe1;
```

```
      e2w_wrpipe1 <= 1;
      e2w_readpipe1 <= 0;
    end
  load:
   begin
     // load data from data bus to dest
     e2w_destpipe1 <= d2e_destpipe1;
     e2w_datapipe1 <= d2e_datapipe1;
     e2w_wrpipe1 <= 1;
     e2w_readpipe1 <= 0;
   end
  move:
   begin
     // move contents from src1 to dest
     e2w_destpipe1 <= d2e_destpipe1;
     e2w_datapipe1 <= int_src1datapipe1;
     e2w_wrpipe1 <= 1;
     e2w_readpipe1 <= 0;
   end
  read:
   begin
     // read data src1 to output
     e2w_destpipe1 <= reg0;
     e2w_datapipe1 <= int_src1datapipe1;
     e2w_wrpipe1 <= 0;
     e2w_readpipe1 <= 1;
   end
  compare:
   begin
     // compare src1, src2, dest
     // results stored in dest
     if (int_src1datapipe1 >
       int_src2datapipe1)
       e2w_datapipe1[1] <= 1;
     else
       e2w_datapipe1[1] <= 0;
     if (int_src1datapipe1 <
       int_src2datapipe1)
       e2w_datapipe1[2] <= 1;
     else
       e2w_datapipe1[2] <= 0;
     if (int_src1datapipe1 <=
       int_src2datapipe1)
       e2w_datapipe1[3] <= 1;
     else
       e2w_datapipe1[3] <= 0;
     if (int_src1datapipe1 >=
       int_src2datapipe1)
       e2w_datapipe1[4] <= 1;
     else
       e2w_datapipe1[4] <= 0;
     e2w_datapipe1[63:5] <= 0;
     e2w_datapipe1[0] <= 0;
     e2w_destpipe1 <= d2e_destpipe1;
     e2w_wrpipe1 <= 1;
     e2w_readpipe1 <= 0;
   end
  xorinst:
   begin
     // xorinst src1, src2, dest
     e2w_destpipe1 <= d2e_destpipe1;
```

```
        e2w_datapipe1 <= int_src1datapipe1
          ^ int_src2datapipe1;
        e2w_wrpipe1 <= 1;
        e2w_readpipe1 <= 0;
      end
   nandinst:
    begin
        // nandinst src1, src2, dest
        e2w_destpipe1 <= d2e_destpipe1;
        e2w_datapipe1 <=~(int_src1datapipe1
          & int_src2datapipe1);
        e2w_wrpipe1 <= 1;
        e2w_readpipe1 <= 0;
      end
   norinst:
    begin
        // norinst src1, src2, dest
        e2w_destpipe1 <= d2e_destpipe1;
        e2w_datapipe1 <=~(int_src1datapipe1
          | int_src2datapipe1);
        e2w_wrpipe1 <= 1;
        e2w_readpipe1 <= 0;
      end
   notinst:
    begin
        // notinst src1, dest
        e2w_destpipe1 <= d2e_destpipe1;
        e2w_datapipe1 <=~int_src1datapipe1;
        e2w_wrpipe1 <= 1;
        e2w_readpipe1 <= 0;
      end
   shiftleft:
    begin
     // shiftleft src1, src2, dest
     e2w_destpipe1 <= d2e_destpipe1;
     case (int_src2datapipe1[3:0])
     4'b0000:
       e2w_datapipe1 <= int_src1datapipe1;
     4'b0001:
       e2w_datapipe1<=(int_src1datapipe1 << 1);
     4'b0010:
       e2w_datapipe1<=(int_src1datapipe1 << 2);
     4'b0011:
       e2w_datapipe1<=(int_src1datapipe1 << 3);
     4'b0100:
       e2w_datapipe1<=(int_src1datapipe1 << 4);
     4'b0101:
       e2w_datapipe1<=(int_src1datapipe1 << 5);
     4'b0110:
       e2w_datapipe1<=(int_src1datapipe1 << 6);
     4'b0111:
       e2w_datapipe1<=(int_src1datapipe1 << 7);
     4'b1000:
       e2w_datapipe1<=(int_src1datapipe1 << 8);
     4'b1001:
       e2w_datapipe1<=(int_src1datapipe1 << 9);
     4'b1010:
       e2w_datapipe1<=(int_src1datapipe1 << 10);
     4'b1011:
       e2w_datapipe1<=(int_src1datapipe1 << 11);
     4'b1100:
       e2w_datapipe1<=(int_src1datapipe1 << 12);
```

```
4'b1101:
  e2w_datapipe1<=(int_src1datapipe1 << 13);
4'b1110:
  e2w_datapipe1<=(int_src1datapipe1 << 14);
4'b1111:
  e2w_datapipe1<=(int_src1datapipe1 << 15);
default:
  e2w_datapipe1<=int_src1datapipe1;
endcase
e2w_wrpipe1 <= 1;
e2w_readpipe1 <= 0;
end
shiftright:
begin
// shiftright src1, src2, dest
e2w_destpipe1 <= d2e_destpipe1;
case (int_src2datapipe1[3:0])
4'b0000:
  e2w_datapipe1 <= int_src1datapipe1;
4'b0001:
  e2w_datapipe1<=(int_src1datapipe1 >> 1);
4'b0010:
  e2w_datapipe1<=(int_src1datapipe1 >> 2);
4'b0011:
  e2w_datapipe1<=(int_src1datapipe1 >> 3);
4'b0100:
  e2w_datapipe1<=(int_src1datapipe1 >> 4);
4'b0101:
  e2w_datapipe1<=(int_src1datapipe1 >> 5);
4'b0110:
  e2w_datapipe1<=(int_src1datapipe1 >> 6);
4'b0111:
  e2w_datapipe1<=(int_src1datapipe1 >> 7);
4'b1000:
  e2w_datapipe1<=(int_src1datapipe1 >> 8);
4'b1001:
  e2w_datapipe1<=(int_src1datapipe1 >> 9);
4'b1010:
  e2w_datapipe1<=(int_src1datapipe1 >> 10);
4'b1011:
  e2w_datapipe1<=(int_src1datapipe1 >> 11);
4'b1100:
  e2w_datapipe1<=(int_src1datapipe1 >> 12);
4'b1101:
  e2w_datapipe1<=(int_src1datapipe1 >> 13);
4'b1110:
  e2w_datapipe1<=(int_src1datapipe1 >> 14);
4'b1111:
  e2w_datapipe1<=(int_src1datapipe1 >> 15);
default:
  e2w_datapipe1 <= int_src1datapipe1;
endcase
e2w_wrpipe1 <= 1;
e2w_readpipe1 <= 0;
end
bshiftleft:
begin
// bshiftleft left src1, src2, dest
e2w_destpipe1 <= d2e_destpipe1;
case (int_src2datapipe1[3:0])
4'b0000:e2w_datapipe1 <= int_src1datapipe1;
4'b0001:e2w_datapipe1<={int_src1datapipe1
  [62:0],int_src1datapipe1[63]};
```

```
4'b0010:e2w_datapipe1<={int_src1datapipe1
 [61:0],int_src1datapipe1[63:62]};
4'b0011:e2w_datapipe1<={int_src1datapipe1
 [60:0],int_src1datapipe1[63:61]};
4'b0100:e2w_datapipe1<={int_src1datapipe1
 [59:0],int_src1datapipe1[63:60]};
4'b0101:e2w_datapipe1<={int_src1datapipe1
 [58:0],int_src1datapipe1[63:59]};
4'b0110:e2w_datapipe1<={int_src1datapipe1
 [57:0],int_src1datapipe1[63:58]};
4'b0111:e2w_datapipe1<={int_src1datapipe1
 [56:0],int_src1datapipe1[63:57]};
4'b1000:e2w_datapipe1<={int_src1datapipe1
 [55:0],int_src1datapipe1[63:56]};
4'b1001:e2w_datapipe1<={int_src1datapipe1
 [54:0],int_src1datapipe1[63:55]};
4'b1010:e2w_datapipe1<={int_src1datapipe1
 [53:0],int_src1datapipe1[63:54]};
4'b1011:e2w_datapipe1<={int_src1datapipe1
 [52:0],int_src1datapipe1[63:53]};
4'b1100:e2w_datapipe1<={int_src1datapipe1
 [51:0],int_src1datapipe1[63:52]};
4'b1101:e2w_datapipe1<={int_src1datapipe1
 [50:0],int_src1datapipe1[63:51]};
4'b1110:e2w_datapipe1<={int_src1datapipe1
 [49:0],int_src1datapipe1[63:50]};
4'b1111:e2w_datapipe1<={int_src1datapipe1
 [48:0],int_src1datapipe1[63:49]};
default:e2w_datapipe1 <= int_src1datapipe1;
endcase
e2w_wrpipe1 <= 1;
e2w_readpipe1 <= 0;
end
bshiftright:
begin
// bshiftright src1, src2, dest
e2w_destpipe1 <= d2e_destpipe1;
 case (int_src2datapipe1[3:0])
4'b0000:e2w_datapipe1 <= int_src1datapipe1;
4'b0001:e2w_datapipe1<={int_src1datapipe1
 [0],int_src1datapipe1[63:1]};
4'b0010:e2w_datapipe1 <= {int_src1datapipe1
 [1:0],int_src1datapipe1[63:2]};
4'b0011:e2w_datapipe1 <= {int_src1datapipe1
 [2:0],int_src1datapipe1[63:3]};
4'b0100:e2w_datapipe1 <= {int_src1datapipe1
 [3:0],int_src1datapipe1[63:4]};
4'b0101:e2w_datapipe1 <= {int_src1datapipe1
 [4:0],int_src1datapipe1[63:5]};
4'b0110:e2w_datapipe1 <= {int_src1datapipe1
 [5:0],int_src1datapipe1[63:6]};
4'b0111:e2w_datapipe1 <= {int_src1datapipe1
 [6:0],int_src1datapipe1[63:7]};
4'b1000:e2w_datapipe1 <= {int_src1datapipe1
 [7:0],int_src1datapipe1[63:8]};
4'b1001:e2w_datapipe1 <= {int_src1datapipe1
 [8:0],int_src1datapipe1[63:9]};
4'b1010:e2w_datapipe1 <= {int_src1datapipe1
 [9:0],int_src1datapipe1[63:10]};
4'b1011:e2w_datapipe1 <= {int_src1datapipe1
 [10:0],int_src1datapipe1[63:11]};
```

```verilog
      4'b1100:e2w_datapipe1 <= {int_src1datapipe1
        [11:0],int_src1datapipe1[63:12]};
      4'b1101:e2w_datapipe1 <= {int_src1datapipe1
        [12:0],int_src1datapipe1[63:13]};
      4'b1110:e2w_datapipe1 <= {int_src1datapipe1
        [13:0],int_src1datapipe1[63:14]};
      4'b1111:e2w_datapipe1 <= {int_src1datapipe1
        [14:0],int_src1datapipe1[63:15]};
      default:e2w_datapipe1 <= int_src1datapipe1;
      endcase
      e2w_wrpipe1 <= 1;
      e2w_readpipe1 <= 0;
      end
   default:
    begin
      // default
      e2w_destpipe1 <= reg0;
      e2w_datapipe1 <= 0;
      e2w_wrpipe1 <= 0;
      e2w_readpipe1 <= 0;
     end
 endcase

 // execute for operation 2 pipe2
 case (d2e_instpipe2)
   nop:
    begin
      // in noop, all default to zero
      e2w_destpipe2 <= reg0;
      e2w_datapipe2 <= 0;
      e2w_wrpipe2 <= 0;
      e2w_readpipe2 <= 0;
     end
   add:
    begin
      // src1 + src2 -> dest
      e2w_destpipe2 <= d2e_destpipe2;
      e2w_datapipe2 <= int_src1datapipe2
        + int_src2datapipe2;
      e2w_wrpipe2 <= 1;
      e2w_readpipe2 <= 0;
     end
   sub:
    begin
      // src1 - src2 -> dest
      e2w_destpipe2 <= d2e_destpipe2;
      e2w_datapipe2 <= int_src1datapipe2
        - int_src2datapipe2;
      e2w_wrpipe2 <= 1;
      e2w_readpipe2 <= 0;
     end
   mul:
    begin
      // src1 x src2 -> dest
      // only 32 bits considered
      e2w_destpipe2 <= d2e_destpipe2;
      e2w_datapipe2 <= int_src1datapipe2
        * int_src2datapipe2;
      e2w_wrpipe2 <= 1;
      e2w_readpipe2 <= 0;
     end
```

```
load:
 begin
  // load data from data bus to dest
  e2w_destpipe2 <= d2e_destpipe2;
  e2w_datapipe2 <= d2e_datapipe2;
  e2w_wrpipe2 <= 1;
  e2w_readpipe2 <= 0;
 end
move:
 begin
  // move contents from src1 to dest
  e2w_destpipe2 <= d2e_destpipe2;
  e2w_datapipe2 <= int_src1datapipe2;
  e2w_wrpipe2 <= 1;
  e2w_readpipe2 <= 0;
 end
read:
 begin
  // read data src1 to output
  e2w_destpipe2 <= reg0;
  e2w_datapipe2 <= int_src1datapipe2;
  e2w_wrpipe2 <= 0;
  e2w_readpipe2 <= 1;
 end
compare:
 begin
  // compare src1, src2, dest
  // results stored in dest
  if (int_src1datapipe2 >
    int_src2datapipe2)
    e2w_datapipe2[1] <= 1;
  else
    e2w_datapipe2[1] <= 0;
  if (int_src1datapipe2 <
    int_src2datapipe2)
    e2w_datapipe2[2] <= 1;
  else
    e2w_datapipe2[2] <= 0;
  if (int_src1datapipe2 <=
    int_src2datapipe2)
    e2w_datapipe2[3] <= 1;
  else
    e2w_datapipe2[3] <= 0;
  if (int_src1datapipe2 >=
    int_src2datapipe2)
    e2w_datapipe2[4] <= 1;
  else
    e2w_datapipe2[4] <= 0;
  e2w_datapipe2[63:5] <= 0;
  e2w_datapipe2[0] <= 0;

  e2w_destpipe2 <= d2e_destpipe2;
  e2w_wrpipe2 <= 1;
  e2w_readpipe2 <= 0;
 end
xorinst:
 begin
  // xorinst src1, src2, dest
  e2w_destpipe2 <= d2e_destpipe2;
  e2w_datapipe2 <= int_src1datapipe2
    ^ int_src2datapipe2;
  e2w_wrpipe2 <= 1;
  e2w_readpipe2 <= 0;
 end
```

```
nandinst:
 begin
   // nandinst src1, src2, dest
   e2w_destpipe2 <= d2e_destpipe2;
   e2w_datapipe2 <=~(int_src1datapipe2
     & int_src2datapipe2);
   e2w_wrpipe2 <= 1;
   e2w_readpipe2 <= 0;
 end
norinst:
 begin
   // norinst src1, src2, dest
   e2w_destpipe2 <= d2e_destpipe2;
   e2w_datapipe2 <=~(int_src1datapipe2
     | int_src2datapipe2);
   e2w_wrpipe2 <= 1;
   e2w_readpipe2 <= 0;
 end
notinst:
 begin
   // notinst src1, dest
   e2w_destpipe2 <= d2e_destpipe2;
   e2w_datapipe2 <=~int_src1datapipe2;
   e2w_wrpipe2 <= 1;
   e2w_readpipe2 <= 0;
 end
shiftleft:
 begin
 // shiftleft src1, src2, dest
 e2w_destpipe2 <= d2e_destpipe2;
 case (int_src2datapipe2[3:0])
 4'b0000:
   e2w_datapipe2<=int_src1datapipe2;
 4'b0001:
   e2w_datapipe2<=(int_src1datapipe2 << 1);
 4'b0010:
   e2w_datapipe2<=(int_src1datapipe2 << 2);
 4'b0011:
   e2w_datapipe2<=(int_src1datapipe2 << 3);
 4'b0100:
   e2w_datapipe2<=(int_src1datapipe2 << 4);
 4'b0101:
   e2w_datapipe2<=(int_src1datapipe2 << 5);
 4'b0110:
   e2w_datapipe2<=(int_src1datapipe2 << 6);
 4'b0111:
   e2w_datapipe2<=(int_src1datapipe2 << 7);
 4'b1000:
   e2w_datapipe2<=(int_src1datapipe2 << 8);
 4'b1001:
   e2w_datapipe2<=(int_src1datapipe2 << 9);
 4'b1010:
   e2w_datapipe2<=(int_src1datapipe2 << 10);
 4'b1011:
   e2w_datapipe2<=(int_src1datapipe2 << 11);
 4'b1100:
   e2w_datapipe2<=(int_src1datapipe2 << 12);
 4'b1101:
   e2w_datapipe2<=(int_src1datapipe2 << 13);
 4'b1110:
   e2w_datapipe2<=(int_src1datapipe2 << 14);
```

```
4'b1111:
  e2w_datapipe2<=(int_src1datapipe2 << 15);
default:
  e2w_datapipe2 <= int_src1datapipe2;
endcase
e2w_wrpipe2 <= 1;
e2w_readpipe2 <= 0;
end
shiftright:
begin
// shiftright src1, src2, dest
e2w_destpipe2 <= d2e_destpipe2;
case (int_src2datapipe2[3:0])
4'b0000:
  e2w_datapipe2 <= int_src1datapipe2;
4'b0001:
  e2w_datapipe2<=(int_src1datapipe2 >> 1);
4'b0010:
  e2w_datapipe2<=(int_src1datapipe2 >> 2);
4'b0011:
  e2w_datapipe2<=(int_src1datapipe2 >> 3);
4'b0100:
  e2w_datapipe2<=(int_src1datapipe2 >> 4);
4'b0101:
  e2w_datapipe2<=(int_src1datapipe2 >> 5);
4'b0110:
  e2w_datapipe2<=(int_src1datapipe2 >> 6);
4'b0111:
  e2w_datapipe2<=(int_src1datapipe2 >> 7);
4'b1000:
  e2w_datapipe2<=(int_src1datapipe2 >> 8);
4'b1001:
  e2w_datapipe2<=(int_src1datapipe2 >> 9);
4'b1010:
  e2w_datapipe2<=(int_src1datapipe2 >> 10);
4'b1011:
  e2w_datapipe2<=(int_src1datapipe2 >> 11);
4'b1100:
  e2w_datapipe2<=(int_src1datapipe2 >> 12);
4'b1101:
  e2w_datapipe2<=(int_src1datapipe2 >> 13);
4'b1110:
  e2w_datapipe2<=(int_src1datapipe2 >> 14);
4'b1111:
  e2w_datapipe2<=(int_src1datapipe2 >> 15);
default:
  e2w_datapipe2 <= int_src1datapipe2;
endcase
e2w_wrpipe2 <= 1;
e2w_readpipe2 <= 0;
end
bshiftleft:
begin
// bshiftleft left src1, src2, dest
e2w_destpipe2 <= d2e_destpipe2;
case (int_src2datapipe2[3:0])
4'b0000:e2w_datapipe2 <= int_src1datapipe2;
4'b0001:e2w_datapipe2 <= {int_src1datapipe2
  [62:0],int_src1datapipe2[63]};
4'b0010:e2w_datapipe2 <= {int_src1datapipe2
  [61:0],int_src1datapipe2[63:62]};
```

```verilog
4'b0011:e2w_datapipe2 <= {int_src1datapipe2
   [60:0],int_src1datapipe2[63:61]};
4'b0100:e2w_datapipe2 <= {int_src1datapipe2
   [59:0],int_src1datapipe2[63:60]};
4'b0101:e2w_datapipe2 <= {int_src1datapipe2
   [58:0],int_src1datapipe2[63:59]};
4'b0110:e2w_datapipe2 <= {int_src1datapipe2
   [57:0],int_src1datapipe2[63:58]};
4'b0111:e2w_datapipe2 <= {int_src1datapipe2
   [56:0],int_src1datapipe2[63:57]};
4'b1000:e2w_datapipe2 <= {int_src1datapipe2
   [55:0],int_src1datapipe2[63:56]};
4'b1001:e2w_datapipe2 <= {int_src1datapipe2
   [54:0],int_src1datapipe2[63:55]};
4'b1010:e2w_datapipe2 <= {int_src1datapipe2
   [53:0],int_src1datapipe2[63:54]};
4'b1011:e2w_datapipe2 <= {int_src1datapipe2
   [52:0],int_src1datapipe2[63:53]};
4'b1100:e2w_datapipe2 <= {int_src1datapipe2
   [51:0],int_src1datapipe2[63:52]};
4'b1101:e2w_datapipe2 <= {int_src1datapipe2
   [50:0],int_src1datapipe2[63:51]};
4'b1110:e2w_datapipe2 <= {int_src1datapipe2
   [49:0],int_src1datapipe2[63:50]};
4'b1111:e2w_datapipe2 <= {int_src1datapipe2
   [48:0],int_src1datapipe2[63:49]};
default:e2w_datapipe2 <= int_src1datapipe2;
endcase
e2w_wrpipe2 <= 1;
e2w_readpipe2 <= 0;
end
bshiftright:
begin
// bshiftright src1, src2, dest
e2w_destpipe2 <= d2e_destpipe2;
case (int_src2datapipe2[3:0])
4'b0000:e2w_datapipe2 <= int_src1datapipe2;
4'b0001:e2w_datapipe2 <= {int_src1datapipe2
   [0],int_src1datapipe2[63:1]};
4'b0010:e2w_datapipe2 <= {int_src1datapipe2
   [1:0],int_src1datapipe2[63:2]};
4'b0011:e2w_datapipe2 <= {int_src1datapipe2
   [2:0],int_src1datapipe2[63:3]};
4'b0100:e2w_datapipe2 <= {int_src1datapipe2
   [3:0],int_src1datapipe2[63:4]};
4'b0101:e2w_datapipe2 <= {int_src1datapipe2
   [4:0],int_src1datapipe2[63:5]};
4'b0110:e2w_datapipe2 <= {int_src1datapipe2
   [5:0],int_src1datapipe2[63:6]};
4'b0111:e2w_datapipe2 <= {int_src1datapipe2
   [6:0],int_src1datapipe2[63:7]};
4'b1000:e2w_datapipe2 <= {int_src1datapipe2
   [7:0],int_src1datapipe2[63:8]};
4'b1001:e2w_datapipe2 <= {int_src1datapipe2
   [8:0],int_src1datapipe2[63:9]};
4'b1010:e2w_datapipe2 <= {int_src1datapipe2
   [9:0],int_src1datapipe2[63:10]};
4'b1011:e2w_datapipe2 <= {int_src1datapipe2
   [10:0],int_src1datapipe2[63:11]};
4'b1100:e2w_datapipe2 <= {int_src1datapipe2
   [11:0],int_src1datapipe2[63:12]};
```

```verilog
      4'b1101:e2w_datapipe2 <= {int_src1datapipe2
        [12:0],int_src1datapipe2[63:13]};
      4'b1110:e2w_datapipe2 <= {int_src1datapipe2
        [13:0],int_src1datapipe2[63:14]};
      4'b1111:e2w_datapipe2 <= {int_src1datapipe2
        [14:0],int_src1datapipe2[63:15]};
      default:e2w_datapipe2 <= int_src1datapipe2;
      endcase
      e2w_wrpipe2 <= 1;
      e2w_readpipe2 <= 0;
      end
   default:
    begin
     // default
     e2w_destpipe2 <= reg0;
     e2w_datapipe2 <= 0;
     e2w_wrpipe2 <= 0;
     e2w_readpipe2 <= 0;
    end
 endcase
// execute for operation 3 pipe3
case (d2e_instpipe3)
  nop:
    begin
      // in noop, all default to zero
      e2w_destpipe3 <= reg0;
      e2w_datapipe3 <= 0;
      e2w_wrpipe3 <= 0;
      e2w_readpipe3 <= 0;
    end
  add:
    begin
      // src1 + src2 -> dest
      e2w_destpipe3 <= d2e_destpipe3;
      e2w_datapipe3 <= int_src1datapipe3
        + int_src2datapipe3;
      e2w_wrpipe3 <= 1;
      e2w_readpipe3 <= 0;
    end
  sub:
    begin
      // src1 - src2 -> dest
      // if borrow occurs, it is ignored
      e2w_destpipe3 <= d2e_destpipe3;
      e2w_datapipe3 <= int_src1datapipe3
        - int_src2datapipe3;
      e2w_wrpipe3 <= 1;
      e2w_readpipe3 <= 0;
    end
  mul:
    begin
      // src1 x src2 -> dest
      // only 32 bits considered
      e2w_destpipe3 <= d2e_destpipe3;
      e2w_datapipe3 <= int_src1datapipe3
        * int_src2datapipe3;
      e2w_wrpipe3 <= 1;
      e2w_readpipe3 <= 0;
    end
```

```
load:
  begin
    // load data from data bus to dest
    e2w_destpipe3 <= d2e_destpipe3;
    e2w_datapipe3 <= d2e_datapipe3;
    e2w_wrpipe3 <= 1;
    e2w_readpipe3 <= 0;
  end
move:
  begin
    // move contents from src1 to dest
    e2w_destpipe3 <= d2e_destpipe3;
    e2w_datapipe3 <= int_src1datapipe3;
    e2w_wrpipe3 <= 1;
    e2w_readpipe3 <= 0;
  end
read:
  begin
    // read data src1 to output
    e2w_destpipe3 <= reg0;
    e2w_datapipe3 <= int_src1datapipe3;
    e2w_wrpipe3 <= 0;
    e2w_readpipe3 <= 1;
  end
compare:
  begin
    // compare src1, src2, dest
    // results stored in dest
    if (int_src1datapipe3 >
      int_src2datapipe3)
      e2w_datapipe3[1] <= 1;
    else
      e2w_datapipe3[1] <= 0;
    if (int_src1datapipe3 <
      int_src2datapipe3)
      e2w_datapipe3[2] <= 1;
    else
      e2w_datapipe3[2] <= 0;
    if (int_src1datapipe3 <=
      int_src2datapipe3)
      e2w_datapipe3[3] <= 1;
    else
      e2w_datapipe3[3] <= 0;
    if (int_src1datapipe3 >=
      int_src2datapipe3)
      e2w_datapipe3[4] <= 1;
    else
      e2w_datapipe3[4] <= 0;
    e2w_datapipe3[63:5] <= 0;
    e2w_datapipe3[0] <= 0;
    e2w_destpipe3 <= d2e_destpipe3;
    e2w_wrpipe3 <= 1;
    e2w_readpipe3 <= 0;
  end
xorinst:
  begin
    // xorinst src1, src2, dest
    e2w_destpipe3 <= d2e_destpipe3;
    e2w_datapipe3 <= int_src1datapipe3
      ^ int_src2datapipe3;
    e2w_wrpipe3 <= 1;
    e2w_readpipe3 <= 0;
  end
```

```
nandinst:
 begin
   // nandinst src1, src2, dest
   e2w_destpipe3 <= d2e_destpipe3;
   e2w_datapipe3 <=~(int_src1datapipe3
     & int_src2datapipe3);
   e2w_wrpipe3 <= 1;
   e2w_readpipe3 <= 0;
 end
norinst:
 begin
   // norinst src1, src2, dest
   e2w_destpipe3 <= d2e_destpipe3;
   e2w_datapipe3 <=~(int_src1datapipe3
    | int_src2datapipe3);
   e2w_wrpipe3 <= 1;
   e2w_readpipe3 <= 0;
 end
notinst:
 begin
   // notinst src1, dest
   e2w_destpipe3 <= d2e_destpipe3;
   e2w_datapipe3 <=~int_src1datapipe3;
   e2w_wrpipe3 <= 1;
   e2w_readpipe3 <= 0;
 end
shiftleft:
 begin
 // shiftleft src1, src2, dest
 e2w_destpipe3 <= d2e_destpipe3;
 case (int_src2datapipe3[3:0])
 4'b0000:
   e2w_datapipe3 <= int_src1datapipe3;
 4'b0001:
   e2w_datapipe3 <=(int_src1datapipe3 << 1);
 4'b0010:
   e2w_datapipe3 <=(int_src1datapipe3 << 2);
 4'b0011:
   e2w_datapipe3 <=(int_src1datapipe3 << 3);
 4'b0100:
   e2w_datapipe3 <=(int_src1datapipe3 << 4);
 4'b0101:
   e2w_datapipe3 <=(int_src1datapipe3 << 5);
 4'b0110:
   e2w_datapipe3 <=(int_src1datapipe3 << 6);
 4'b0111:
   e2w_datapipe3 <=(int_src1datapipe3 << 7);
 4'b1000:
   e2w_datapipe3 <=(int_src1datapipe3 << 8);
 4'b1001:
   e2w_datapipe3 <=(int_src1datapipe3 << 9);
 4'b1010:
   e2w_datapipe3<=(int_src1datapipe3 << 10);
 4'b1011:
   e2w_datapipe3<=(int_src1datapipe3 << 11);
 4'b1100:
   e2w_datapipe3<=(int_src1datapipe3 << 12);
 4'b1101:
   e2w_datapipe3<=(int_src1datapipe3 << 13);
 4'b1110:
   e2w_datapipe3<=(int_src1datapipe3 << 14);
 4'b1111:
   e2w_datapipe3<=(int_src1datapipe3 << 15);
 default:
   e2w_datapipe3 <= int_src1datapipe3;
 endcase
```

```
    e2w_wrpipe3 <= 1;
    e2w_readpipe3 <= 0;
    end
shiftright:
 begin
 // shiftright src1, src2, dest
 e2w_destpipe3 <= d2e_destpipe3;
 case (int_src2datapipe3[3:0])
 4'b0000:
   e2w_datapipe3 <= int_src1datapipe3;
 4'b0001:
   e2w_datapipe3 <=(int_src1datapipe3 >> 1);
 4'b0010:
   e2w_datapipe3 <=(int_src1datapipe3 >> 2);
 4'b0011:
   e2w_datapipe3 <=(int_src1datapipe3 >> 3);
 4'b0100:
   e2w_datapipe3 <=(int_src1datapipe3 >> 4);
 4'b0101:
   e2w_datapipe3 <=(int_src1datapipe3 >> 5);
 4'b0110:
   e2w_datapipe3 <=(int_src1datapipe3 >> 6);
 4'b0111:
   e2w_datapipe3 <=(int_src1datapipe3 >> 7);
 4'b1000:
   e2w_datapipe3 <=(int_src1datapipe3 >> 8);
 4'b1001:
   e2w_datapipe3 <=(int_src1datapipe3 >> 9);
 4'b1010:
   e2w_datapipe3<=(int_src1datapipe3 >> 10);
 4'b1011:
   e2w_datapipe3<=(int_src1datapipe3 >> 11);
 4'b1100:
   e2w_datapipe3<=(int_src1datapipe3 >> 12);
 4'b1101:
   e2w_datapipe3<=(int_src1datapipe3 >> 13);
 4'b1110:
   e2w_datapipe3<=(int_src1datapipe3 >> 14);
 4'b1111:
   e2w_datapipe3<=(int_src1datapipe3 >> 15);
 default:
   e2w_datapipe3 <= int_src1datapipe3;
 endcase
 e2w_wrpipe3 <= 1;
 e2w_readpipe3 <= 0;
 end
bshiftleft:
 begin
 // bshiftleft left src1, src2, dest
 e2w_destpipe3 <= d2e_destpipe3;
 case (int_src2datapipe3[3:0])
 4'b0000:e2w_datapipe3 <= int_src1datapipe3;
 4'b0001:e2w_datapipe3 <= {int_src1datapipe3
   [62:0],int_src1datapipe3[63]};
 4'b0010:e2w_datapipe3 <= {int_src1datapipe3
   [61:0],int_src1datapipe3[63:62]};
 4'b0011:e2w_datapipe3 <= {int_src1datapipe3
   [60:0],int_src1datapipe3[63:61]};
 4'b0100:e2w_datapipe3 <= {int_src1datapipe3
   [59:0],int_src1datapipe3[63:60]};
 4'b0101:e2w_datapipe3 <= {int_src1datapipe3
   [58:0],int_src1datapipe3[63:59]};
 4'b0110:e2w_datapipe3 <= {int_src1datapipe3
   [57:0],int_src1datapipe3[63:58]};
```

```
4'b0111:e2w_datapipe3 <= {int_src1datapipe3
   [56:0],int_src1datapipe3[63:57]};
4'b1000:e2w_datapipe3 <= {int_src1datapipe3
   [55:0],int_src1datapipe3[63:56]};
4'b1001:e2w_datapipe3 <= {int_src1datapipe3
   [54:0],int_src1datapipe3[63:55]};
4'b1010:e2w_datapipe3 <= {int_src1datapipe3
   [53:0],int_src1datapipe3[63:54]};
4'b1011:e2w_datapipe3 <= {int_src1datapipe3
   [52:0],int_src1datapipe3[63:53]};
4'b1100:e2w_datapipe3 <= {int_src1datapipe3
   [51:0],int_src1datapipe3[63:52]};
4'b1101:e2w_datapipe3 <= {int_src1datapipe3
   [50:0],int_src1datapipe3[63:51]};
4'b1110:e2w_datapipe3 <= {int_src1datapipe3
   [49:0],int_src1datapipe3[63:50]};
4'b1111:e2w_datapipe3 <= {int_src1datapipe3
   [48:0],int_src1datapipe3[63:49]};
default:e2w_datapipe3 <= int_src1datapipe3;
endcase
e2w_wrpipe3 <= 1;
e2w_readpipe3 <= 0;
end
bshiftright:
begin
// bshiftright src1, src2, dest
e2w_destpipe3 <= d2e_destpipe3;
case (int_src2datapipe3[3:0])
4'b0000:e2w_datapipe3 <= int_src1datapipe3;
4'b0001:e2w_datapipe3 <= {int_src1datapipe3
   [0],int_src1datapipe3[63:1]};
4'b0010:e2w_datapipe3 <= {int_src1datapipe3
   [1:0],int_src1datapipe3[63:2]};
4'b0011:e2w_datapipe3 <= {int_src1datapipe3
   [2:0],int_src1datapipe3[63:3]};
4'b0100:e2w_datapipe3 <= {int_src1datapipe3
   [3:0],int_src1datapipe3[63:4]};
4'b0101:e2w_datapipe3 <= {int_src1datapipe3
   [4:0],int_src1datapipe3[63:5]};
4'b0110:e2w_datapipe3 <= {int_src1datapipe3
   [5:0],int_src1datapipe3[63:6]};
4'b0111:e2w_datapipe3 <= {int_src1datapipe3
   [6:0],int_src1datapipe3[63:7]};
4'b1000:e2w_datapipe3 <= {int_src1datapipe3
   [7:0],int_src1datapipe3[63:8]};
4'b1001:e2w_datapipe3 <= {int_src1datapipe3
   [8:0],int_src1datapipe3[63:9]};
4'b1010:e2w_datapipe3 <= {int_src1datapipe3
   [9:0],int_src1datapipe3[63:10]};
4'b1011:e2w_datapipe3 <= {int_src1datapipe3
   [10:0],int_src1datapipe3[63:11]};
4'b1100:e2w_datapipe3 <= {int_src1datapipe3
   [11:0],int_src1datapipe3[63:12]};
4'b1101:e2w_datapipe3 <= {int_src1datapipe3
   [12:0],int_src1datapipe3[63:13]};
4'b1110:e2w_datapipe3 <= {int_src1datapipe3
   [13:0],int_src1datapipe3[63:14]};
4'b1111:e2w_datapipe3 <= {int_src1datapipe3
   [14:0],int_src1datapipe3[63:15]};
default:e2w_datapipe3 <= int_src1datapipe3;
endcase
e2w_wrpipe3 <= 1;
e2w_readpipe3 <= 0;
end
```

```
          default:
            begin
              // default
              e2w_destpipe3 <= reg0;
              e2w_datapipe3 <= 0;
              e2w_wrpipe3 <= 0;
              e2w_readpipe3 <= 0;
            end
        endcase
        if (((d2e_instpipe3 == compare) & (int_src1datapipe3
        == int_src2datapipe3)) | ((d2e_instpipe2 == compare)
        & (int_src1datapipe2 == int_src2datapipe2)) |
        ((d2e_instpipe1 == compare) & (int_src1datapipe1 ==
        int_src2datapipe1)))
          begin
            preflush <= 1;
            jump <= 1;
          end
        else
          begin
            preflush <= 0;
            jump <= 0;
          end
    end
end
// flush needs to be delayed 1 clock cycle to ensure adequate time for
// writeback to write the necessary data into registers in register file

always @ (posedge clock or posedge reset)
begin
  if (reset)
    begin
      flush <= 0;
    end
  else
    begin
      flush <= preflush;
    end
end

endmodule
```

3.2.5 Module *writeback* RTL Code

The writeback module is the last stage within the VLIW micro-processor. Its functionality is to write the results of executed operations into the register file. Table 3.10 shows the interface signals for the writeback module and its interface signal functionality. Figure 3.15 shows the interface signal diagram of the writeback module.

Based on the interface signals shown in Table 3.10 with the signal functionality, the RTL verilog code for the writeback module is shown in Example 3.10.

Example 3.10 RTL Verilog Code of writeback Module

```
module writeback (clock, reset, flush,
e2w_destpipe1, e2w_destpipe2, e2w_destpipe3,
e2w_datapipe1, e2w_datapipe2, e2w_datapipe3,
```

TABLE 3.10   Interface Signals of *writeback* Module

Signal Name	Input/ Output	Bits	Description
clock	Input	1	Input clock pin. The VLIW microprocessor is active on rising edge of clock.
reset	Input	1	Input reset pin. Reset is asynchronous and active high.
flush	Input	1	This is a global signal that flushes all the modules, indicating that a branch is to occur.
e2w_destpipe1	Input	4	Represents the destination register for operation 1.
e2w_destpipe2	Input	4	Represents the destination register for operation 2.
e2w_destpipe3	Input	4	Represents the destination register for operation 3.
e2w_datapipe1	Input	64	Represents the data for operation 1.
e2w_datapipe2	Input	64	Represents the data for operation 2.
e2w_datapipe3	Input	64	Represents the data for operation 3.
e2w_wrpipe1	Input	1	Represents the write signal from execute module to writeback module. This signal is passed from writeback module to register file module. It indicates the contents of w2re_datapipe1 to be stored into register specified by w2re_destpipe1.
e2w_wrpipe2	Input	1	Represents the write signal from execute module to writeback module. This signal is passed from writeback module to register file module. It indicates the contents of w2re_datapipe2 to be stored into register specified by w2re_destpipe2.
e2w_wrpipe3	Input	1	Represents the write signal from execute module to writeback module. This signal is passed from writeback module to register file module. It indicates the contents of w2re_datapipe3 to be stored into register specified by w2re_destpipe3.
e2w_readpipe1	Input	1	This signal indicates to the writeback module that the data on e2w_datapipe1 is to be read out of the VLIW microprocessor, through the output port readdatapipe1.

(Continued)

TABLE 3.10   Interface Signals of *writeback* Module (*Continued*)

Signal Name	Input/ Output	Bits	Description
e2w_readpipe2	Input	1	This signal indicates to the writeback module that the data on e2w_datapipe2 is to be read out of the VLIW microprocessor, through the output port readdatapipe2.
e2w_readpipe3	Input	1	This signal indicates to the writeback module that the data on e2w_datapipe3 is to be read out of the VLIW microprocessor, through the output port readdatapipe3.
w2r_wrpipe1	Output	1	Represents the write signal to register file module. When this signal is logic 1, contents of w2re_datapipe1 is stored into register specified by w2re_destpipe1.
w2r_wrpipe2	Output	1	Represents the write signal to register file module. When this signal is logic 1, contents of w2re_datapipe2 is stored into register specified by w2re_destpipe2.
w2r_wrpipe3	Output	1	Represents the write signal to register file module. When this signal is logic 1, contents of w2re_datapipe3 is stored into register specified by w2re_destpipe3.
w2re_destpipe1	Output	4	Represents the destination register of operation 1.
w2re_destpipe2	Output	4	Represents the destination register of operation 2.
w2re_destpipe3	Output	4	Represents the destination register of operation 3.
w2re_datapipe1	Output	64	Represents the 64-bit result of operation 1 executed by execute module. This data are written into the register file module if signal w2r_wrpipe1 is at logic 1.
w2re_datapipe2	Output	64	Represents the 64-bit result of operation 2 executed by execute module. This data are written into the register file module if signal w2r_wrpipe2 is at logic 1.
w2re_datapipe3	Output	64	Represents the 64-bit result of operation 3 executed by execute module. This data are written into the register file module if signal w2r_wrpipe3 is at logic 1.

TABLE 3.10   Interface Signals of *writeback* Module (*Continued*)

Signal Name	Input/ Output	Bits	Description
readdatapipe1	Output	64	Represents the 64-bit data read out of the VLIW microprocessor for operation1.
readdatapipe2	Output	64	Represents the 64-bit data read out of the VLIW microprocessor for operation2.
readdatapipe3	Output	64	Represents the 64-bit data read out of the VLIW microprocessor for operation3.
readdatavalid	Output	1	Represents a data valid condition on the output port of readdatapipe1, readdatapipe2, and readdatapipe3.

```
e2w_wrpipe1, e2w_wrpipe2, e2w_wrpipe3,
e2w_readpipe1, e2w_readpipe2, e2w_readpipe3,
w2r_wrpipe1, w2r_wrpipe2, w2r_wrpipe3,
w2re_destpipe1, w2re_destpipe2, w2re_destpipe3,
w2re_datapipe1, w2re_datapipe2, w2re_datapipe3,
readdatapipe1, readdatapipe2, readdatapipe3,
readdatavalid);

input clock, reset, flush;
input [3:0]e2w_destpipe1, e2w_destpipe2, e2w_destpipe3;
input [63:0] e2w_datapipe1, e2w_datapipe2, e2w_datapipe3;
input e2w_wrpipe1, e2w_wrpipe2, e2w_wrpipe3;
input e2w_readpipe1, e2w_readpipe2, e2w_readpipe3;
output w2r_wrpipe1, w2r_wrpipe2, w2r_wrpipe3;
output [3:0] w2re_destpipe1, w2re_destpipe2, w2re_destpipe3;
output [63:0] w2re_datapipe1, w2re_datapipe2, w2re_datapipe3;
output [63:0] readdatapipe1, readdatapipe2, readdatapipe3;
output readdatavalid;
```

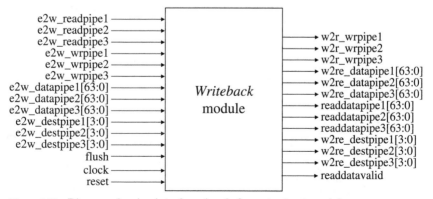

Figure 3.15 Diagram showing interface signals for writeback module.

```verilog
reg w2r_wrpipe1, w2r_wrpipe2, w2r_wrpipe3;
reg [3:0] w2re_destpipe1, w2re_destpipe2, w2re_destpipe3;
reg [63:0] w2re_datapipe1, w2re_datapipe2, w2re_datapipe3;
reg [63:0] readdatapipe1, readdatapipe2, readdatapipe3;
reg readdatavalid;

// include the file that declares the parameter declaration for
// register names and also instruction operations
`include "regname.v"

always @ (posedge clock or posedge reset)
begin
  if (reset)
  begin
   w2r_wrpipe1 <= 0;
   w2r_wrpipe2 <= 0;
   w2r_wrpipe3 <= 0;
   w2re_destpipe1 <= reg0;
   w2re_destpipe2 <= reg0;
   w2re_destpipe3 <= reg0;
   w2re_datapipe1 <= 0;
   w2re_datapipe2 <= 0;
   w2re_datapipe3 <= 0;
   readdatapipe1 <= 0;
   readdatapipe2 <= 0;
   readdatapipe3 <= 0;
   readdatavalid <= 0;
  end
  else // positive edge of clock detected
  begin
   if (~flush)
   begin
     if (e2w_readpipe1)
       begin
         readdatapipe1 <= e2w_datapipe1;
       end
     else
       begin
         readdatapipe1 <= 0;
       end

     if (e2w_readpipe2)
       begin
         readdatapipe2 <= e2w_datapipe2;
       end
     else
       begin
         readdatapipe2 <= 0;
       end
     if (e2w_readpipe3)
       begin
         readdatapipe3 <= e2w_datapipe3;
       end
     else
       begin
         readdatapipe3 <= 0;
       end
     readdatavalid <= e2w_readpipe1 | e2w_readpipe2 |
         e2w_readpipe3;

     w2r_wrpipe1 <= e2w_wrpipe1;
     w2r_wrpipe2 <= e2w_wrpipe2;
     w2r_wrpipe3 <= e2w_wrpipe3;
```

```
        w2re_destpipe1 <= e2w_destpipe1;
        w2re_destpipe2 <= e2w_destpipe2;
        w2re_destpipe3 <= e2w_destpipe3;

        w2re_datapipe1 <= e2w_datapipe1;
        w2re_datapipe2 <= e2w_datapipe2;
        w2re_datapipe3 <= e2w_datapipe3;
      end
      else // flush
      begin
        w2r_wrpipe1 <= 0;
        w2r_wrpipe2 <= 0;
        w2r_wrpipe3 <= 0;
        w2re_destpipe1 <= reg0;
        w2re_destpipe2 <= reg0;
        w2re_destpipe3 <= reg0;
        w2re_datapipe1 <= 0;
        w2re_datapipe2 <= 0;
        w2re_datapipe3 <= 0;
        readdatapipe1 <= 0;
        readdatapipe2 <= 0;
        readdatapipe3 <= 0;
        readdatavalid <= 0;
      end
    end
  end
end
endmodule
```

3.2.6   Module vliwtop RTL Code

The vliwtop module is the top level module of the VLIW microprocessor. It is a top level instantiation of the five modules of fetch, decode, execute, writeback, and register file. The top level interface signals for the VLIW microprocessor and its interface signal functionality are shown in Table 2.20. Figure 3.16 shows the interface signal diagram.

The module vliwtop is integrated with periphery modules as shown in Figure 2.5. Example 3.11 shows the RTL verilog code for the module vliwtop.

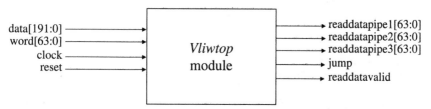

Figure 3.16   Diagram showing interface signals for vliwtop module.

Example 3.11   RTL Verilog Code of `vliwtop` Module

```
module vliw_top (
clock, reset, word, data, readdatapipe1, readdatapipe2, readdatapipe3,
readdatavalid, jump);

input clock, reset;
input [63:0] word;
input [191:0] data;
output [63:0] readdatapipe1, readdatapipe2, readdatapipe3;
output readdatavalid;
output jump;

wire [63:0] readdatapipe1, readdatapipe2, readdatapipe3;
wire readdatavalid;
wire jump;
wire [3:0] f2dr_instpipe1, f2dr_instpipe2, f2dr_instpipe3;
wire [3:0] f2d_destpipe1, f2d_destpipe2, f2d_destpipe3;
wire [3:0] f2r_src1pipe1, f2r_src1pipe2, f2r_src1pipe3;
wire [3:0] f2r_src2pipe1, f2r_src2pipe2, f2r_src2pipe3;
wire [191:0] f2d_data;
wire [3:0] d2e_instpipe1, d2e_instpipe2, d2e_instpipe3;
wire [3:0] d2e_destpipe1, d2e_destpipe2, d2e_destpipe3;
wire [63:0] d2e_datapipe1, d2e_datapipe2, d2e_datapipe3;
wire [63:0] r2e_src1datapipe1, r2e_src1datapipe2, r2e_src1datapipe3;
wire [63:0] r2e_src2datapipe1, r2e_src2datapipe2, r2e_src2datapipe3;
wire [3:0] e2w_destpipe1, e2w_destpipe2, e2w_destpipe3;
wire [63:0] e2w_datapipe1, e2w_datapipe2, e2w_datapipe3;
wire e2w_wrpipe1, e2w_wrpipe2, e2w_wrpipe3;
wire e2w_readpipe1, e2w_readpipe2, e2w_readpipe3;
wire w2r_wrpipe1, w2r_wrpipe2, w2r_wrpipe3;
wire [3:0] w2re_destpipe1, w2re_destpipe2, w2re_destpipe3;
wire [63:0] w2re_datapipe1, w2re_datapipe2, w2re_datapipe3;
wire [3:0] r2e_src1pipe1, r2e_src1pipe2, r2e_src1pipe3;
wire [3:0] r2e_src2pipe1, r2e_src2pipe2, r2e_src2pipe3;

fetch fetchinst (
.word(word), .data(data),
.f2d_destpipe1(f2d_destpipe1),
.f2d_destpipe2(f2d_destpipe2),
.f2d_destpipe3(f2d_destpipe3),
.f2d_data(f2d_data),
.f2dr_instpipe1(f2dr_instpipe1),
.f2dr_instpipe2(f2dr_instpipe2),
.f2dr_instpipe3(f2dr_instpipe3),
.f2r_src1pipe1(f2r_src1pipe1),
.f2r_src1pipe2(f2r_src1pipe2),
.f2r_src1pipe3(f2r_src1pipe3),
.f2r_src2pipe1(f2r_src2pipe1),
.f2r_src2pipe2(f2r_src2pipe2),
.f2r_src2pipe3(f2r_src2pipe3),
.clock(clock), .reset(reset),
.flush(flush));

decode decodeinst (
.f2d_destpipe1(f2d_destpipe1),
.f2d_destpipe2(f2d_destpipe2),
.f2d_destpipe3(f2d_destpipe3),
.f2d_data(f2d_data),
.f2dr_instpipe1(f2dr_instpipe1),
```

```
.f2dr_instpipe2(f2dr_instpipe2),
.f2dr_instpipe3(f2dr_instpipe3),
.clock(clock),
.reset(reset),
.flush(flush),
.d2e_instpipe1(d2e_instpipe1),
.d2e_instpipe2(d2e_instpipe2),
.d2e_instpipe3(d2e_instpipe3),
.d2e_destpipe1(d2e_destpipe1),
.d2e_destpipe2(d2e_destpipe2),
.d2e_destpipe3(d2e_destpipe3),
.d2e_datapipe1(d2e_datapipe1),
.d2e_datapipe2(d2e_datapipe2),
.d2e_datapipe3(d2e_datapipe3));

execute executeinst (
.clock(clock),
.reset(reset),
.d2e_instpipe1(d2e_instpipe1),
.d2e_instpipe2(d2e_instpipe2),
.d2e_instpipe3(d2e_instpipe3),
.d2e_destpipe1(d2e_destpipe1),
.d2e_destpipe2(d2e_destpipe2),
.d2e_destpipe3(d2e_destpipe3),
.d2e_datapipe1(d2e_datapipe1),
.d2e_datapipe2(d2e_datapipe2),
.d2e_datapipe3(d2e_datapipe3),
.r2e_src1datapipe1(r2e_src1datapipe1),
.r2e_src1datapipe2(r2e_src1datapipe2),
.r2e_src1datapipe3(r2e_src1datapipe3),
.r2e_src2datapipe1(r2e_src2datapipe1),
.r2e_src2datapipe2(r2e_src2datapipe2),
.r2e_src2datapipe3(r2e_src2datapipe3),
.r2e_src1pipe1 (r2e_src1pipe1),
.r2e_src1pipe2 (r2e_src1pipe2),
.r2e_src1pipe3 (r2e_src1pipe3),
.r2e_src2pipe1 (r2e_src2pipe1),
.r2e_src2pipe2 (r2e_src2pipe2),
.r2e_src2pipe3 (r2e_src2pipe3),
.w2re_destpipe1 (w2re_destpipe1),
.w2re_destpipe2 (w2re_destpipe2),
.w2re_destpipe3 (w2re_destpipe3),
.w2re_datapipe1 (w2re_datapipe1),
.w2re_datapipe2 (w2re_datapipe2),
.w2re_datapipe3 (w2re_datapipe3),
.e2w_destpipe1(e2w_destpipe1),
.e2w_destpipe2(e2w_destpipe2),
.e2w_destpipe3(e2w_destpipe3),
.e2w_datapipe1(e2w_datapipe1),
.e2w_datapipe2(e2w_datapipe2),
.e2w_datapipe3(e2w_datapipe3),
.e2w_wrpipe1(e2w_wrpipe1),
.e2w_wrpipe2(e2w_wrpipe2),
.e2w_wrpipe3(e2w_wrpipe3),
.e2w_readpipe1(e2w_readpipe1),
.e2w_readpipe2(e2w_readpipe2),
.e2w_readpipe3(e2w_readpipe3),
.flush(flush),
.jump(jump)
);

writeback writebackinst (
.clock(clock),
.reset(reset),
.flush(flush),
```

Top level execute bypass can be achieved with the following code for src1 on pipe1:

```
wire [63:0]
r2e_src1bypasspipe1 =
((e2w_destpipe1 ==
r2e_src1pipe1)&(d2e_
instpipe1 != load) &
(d2e_instpipe1 != nop)) ?
e2w_datapipe1 :
((e2w_destpipe2 ==
r2e_src1pipe1) &
(d2e_instpipe2 !=
load)&(d2e_instpipe2 !=
nop)) ? e2w_datapipe2 :
((e2w_destpipe3 ==
r2e_src1pipe1) &
(d2e_instpipe3 != load) &
(d2e_instpipe3 != nop)) ?
e2w_datapipe3 :
r2e_src1datapipe1;
```

Replace the execute module connectivity of
```
.r2e_src1datapipe1(r2e_src
1datapipe1)
```
with
```
.r2e_src1datapipe1
(r2e_src1bypasspipe1)
```

Similarly code changes for r2e_src2bypasspipe1 and also for the other pipe2 and pipe3.

```
.e2w_destpipe1(e2w_destpipe1),
.e2w_destpipe2(e2w_destpipe2),
.e2w_destpipe3(e2w_destpipe3),
.e2w_datapipe1(e2w_datapipe1),
.e2w_datapipe2(e2w_datapipe2),
.e2w_datapipe3(e2w_datapipe3),
.e2w_wrpipe1(e2w_wrpipe1),
.e2w_wrpipe2(e2w_wrpipe2),
.e2w_wrpipe3(e2w_wrpipe3),
.e2w_readpipe1(e2w_readpipe1),
.e2w_readpipe2(e2w_readpipe2),
.e2w_readpipe3(e2w_readpipe3),
.w2r_wrpipe1(w2r_wrpipe1),
.w2r_wrpipe2(w2r_wrpipe2),
.w2r_wrpipe3(w2r_wrpipe3),
.w2re_destpipe1(w2re_destpipe1),
.w2re_destpipe2(w2re_destpipe2),
.w2re_destpipe3(w2re_destpipe3),
.w2re_datapipe1(w2re_datapipe1),
.w2re_datapipe2(w2re_datapipe2),
.w2re_datapipe3(w2re_datapipe3),
.readdatapipe1(readdatapipe1),
.readdatapipe2(readdatapipe2),
.readdatapipe3(readdatapipe3),
.readdatavalid(readdatavalid));

registerfile registerfileinst (
.f2r_src1pipe1(f2r_src1pipe1),
.f2r_src1pipe2(f2r_src1pipe2),
.f2r_src1pipe3(f2r_src1pipe3),
.f2r_src2pipe1(f2r_src2pipe1),
.f2r_src2pipe2(f2r_src2pipe2),
.f2r_src2pipe3(f2r_src2pipe3),
.f2dr_instpipe1(f2dr_instpipe1),
.f2dr_instpipe2(f2dr_instpipe2),
.f2dr_instpipe3(f2dr_instpipe3),
.clock(clock),
.flush(flush),
.reset(reset),
.w2re_datapipe1(w2re_datapipe1),
.w2re_datapipe2(w2re_datapipe2),
.w2re_datapipe3(w2re_datapipe3),
.w2r_wrpipe1(w2r_wrpipe1),
.w2r_wrpipe2(w2r_wrpipe2),
.w2r_wrpipe3(w2r_wrpipe3),
.w2re_destpipe1(w2re_destpipe1),
.w2re_destpipe2(w2re_destpipe2),
.w2re_destpipe3(w2re_destpipe3),
.r2e_src1datapipe1(r2e_src1datapipe1),
.r2e_src1datapipe2(r2e_src1datapipe2),
.r2e_src1datapipe3(r2e_src1datapipe3),
.r2e_src2datapipe1(r2e_src2datapipe1),
.r2e_src2datapipe2(r2e_src2datapipe2),
.r2e_src2datapipe3(r2e_src2datapipe3),
.r2e_src1pipe1 (r2e_src1pipe1),
.r2e_src1pipe2 (r2e_src1pipe2),
.r2e_src1pipe3 (r2e_src1pipe3),
.r2e_src2pipe1 (r2e_src2pipe1),
.r2e_src2pipe2 (r2e_src2pipe2),
.r2e_src2pipe3 (r2e_src2pipe3)
);
Endmodule
```

3.3   Testbenches and Simulation

When the RTL code of a design is completed, the next step would be to create testbenches to simulate the design. Testbenches can be in many different forms. Some design engineers use verilog, VHDL, C, systemC, systemVerilog or a mixture of them. Whatever the language used for writing testbench, the end result is the same: creation of testbenches used for simulating the design.

Testbenches is a wrap-around of a design, which allows the testbench to pump in stimulus into the design under test, and monitoring the output of the design. If the output waveforms of the design are not as expected, a bug has occurred. The bug could be in the design or in the testbench.

When a bug is found, the designer must debug the waveforms and decide if it is from the design or the testbench. Either way, the bug must be fixed and simulation is performed again. Only when the output waveform of the design is as expected can the designer proceed to the next phase of the design flow (synthesis).

Figure 3.17 shows how a testbench can wrap around a design for simulation.

Figure 3.18 shows the flow used for simulation of RTL design using testbenches. The testbench and RTL code of the design are simulated using a verilog simulator. There are many verilog simulators available in the market, for example, Modelsim from Mentor Graphics, VCS from Synopsys, NC Verilog from Cadence, and many others. The simulation waveform from the verilog simulation is checked for matches with the expected waveform. If the simulated waveform is not what is expected, the designer will have to modify the RTL code of the design or the testbench, depending on which causes the error. A resimulation is performed, and this action is repeated until the designer is satisfied that the simulated waveform matches what is expected of the design.

Appendix A shows some of the testbenches that are used to verify the VLIW microprocessor and its corresponding simulation waveform.

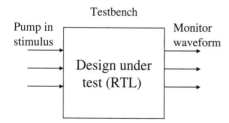

Figure 3.17 Diagram showing a testbench wrapping around a design under test.

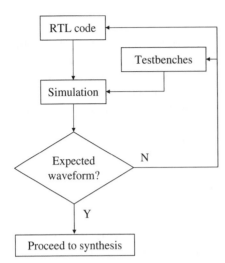

Figure 3.18 Diagram showing flow used for simulation of design.

3.3.1 Creating and Using a Testplan

A design has many features. Each of these features has to be thoroughly simulated in order to ensure that the design is fully functional. As such, it is a good practice to always create a testplan to define the different testbenches that are needed to fully validate the design. The testplan will serve as a useful guide to achieving the targeted verification milestone. Table 3.11 shows an example of a simple testplan that is created for verification of the VLIW microprocessor.

3.3.2 Code Coverage

To fully verify a design, all features of the design need to be simulated. Therefore, each design will have many different testbenches, with

TABLE 3.11 Example of a Simple Testplan for the VLIW Microprocessor

	Week 1	Week 2	Week 3	Week 4	Week 5	Week 6
Basic tests for all 16 instructions	▓					
Tests to check for "jump"		▓				
Tests to check for different combinations of instructions		▓	▓	▓		
Tests to check for all conditions of register bypass					▓	▓
Tests to check for external read						▓

each testbench simulating a certain feature of the design. In a design project, it is common for a design module to be checked for its code coverage to ensure that most portions of the RTL verilog code have been verified.

Code coverage is a method in which a code coverage tool can analyze all the testbenches and the RTL verilog code of the design and provide a report on portions of the RTL code that is not exercised by the testbenches. The more RTL verilog code that is exercised, the better the code coverage. It is common design practice to have at least 95% code coverage for a design module. If the code coverage is less than the targeted rate, more testbenches must be written to verify those parts of the RTL verilog code that are not exercised. It is, however, rather difficult to obtain complete (100%) code coverage, especially for a large design with many lines of RTL verilog code.

Figure 3.19 shows the flow used for code coverage. The code coverage analysis tool analyzes all the different testbenches with the RTL code, and provides a detailed report on portions of the RTL code that are not exercised. More testbenches are written to exercise the unexcersised parts of the RTL code in order to increase the code coverage percentage to a minimum of 95%.

Example 3.12 and Figure 3.20 show examples of a report from a code coverage analysis tool indicating that certain parts of the RTL verilog code of the VLIW microprocessor have not been exercised by the testbenches.

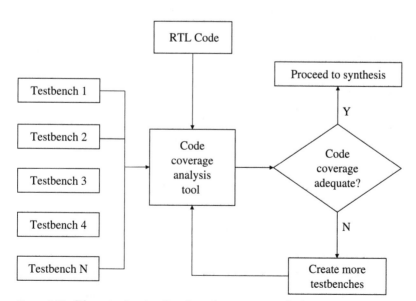

Figure 3.19   Diagram showing flow for code coverage analysis.

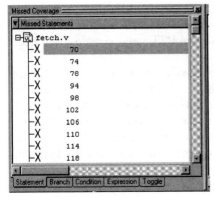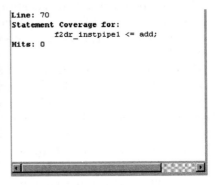

Figure 3.20 Diagram showing an example of missing coverage in coverage analysis report (generated from Mentor Graphics Modelsim Simulator).

Example 3.12 Coverage Report for Code Coverage Analysis (Generated from Mentor Graphics' Modelsim Simulator)

```
# Coverage Report for instance /vliw_top_tb/vliw_top_inst/fetchinst
with # line data
# Statement Coverage:
#    Inst      DU       Stmts     Hits        %        Coverage Enabled
#    ----      ----     ------    -----      ----      ----------------

/vliw_top_tb/
vliw_top_inst/

fetchinst    fetch    248       100        40.3      Stmt
#
#Statement Coverage for instance /vliw_top_tb/vliw_top_inst/fetchinst —
#
#      Line     Stmt     Count    Source
#      -----    -----    -----    -------
   File /project/VLIW/64bit/simulation/fetch.v
       1                         module fetch (
       2                         word, data, clock, reset, flush,
       3                         f2d_data,
       4                         f2d_destpipe1, f2d_destpipe2, f2d_destpipe3,
       5                         f2dr_instpipe1, f2dr_instpipe2, f2dr_instpipe3,
       6                         f2r_src1pipe1, f2r_src1pipe2, f2r_src1pipe3,
       7                         f2r_src2pipe1, f2r_src2pipe2, f2r_src2pipe3
       8                         );
       9
      10                         input clock; // clock input
      11                         input reset; // asynchronous reset active high
      ......
      ......
      29
      30      1        115       always @ (posedge clock or posedge reset)
      31                         begin
      32      1        115       if (reset)
      33                         begin
      34      1        3            f2dr_instpipe1 <= nop;
      35      1        3            f2dr_instpipe2 <= nop;
      36      1        3            f2dr_instpipe3 <= nop;
      37      1        3            f2r_src1pipe1 <= reg0;
```

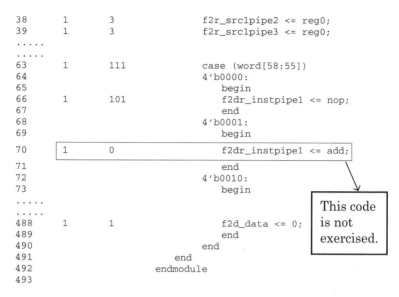

```
38      1       3               f2r_src1pipe2 <= reg0;
39      1       3               f2r_src1pipe3 <= reg0;
.....
.....
63      1       111             case (word[58:55])
64                              4'b0000:
65                                  begin
66      1       101                     f2dr_instpipe1 <= nop;
67                                  end
68                              4'b0001:
69                                  begin
70      1       0                       f2dr_instpipe1 <= add;
71                                  end
72                              4'b0010:
73                                  begin
.....
.....
488     1       1                       f2d_data <= 0;
489                                 end
490                             end
491                     end
492             endmodule
493
```

This code is not exercised.

Referring to Example 3.12, the code coverage is checked for the module fetch in the VLIW microprocessor. The column "count" of the code coverage report indicates if a particular statement of the fetch module is exercised by the testbench. Statements that indicate zero count show that it has not been exercised. The designer will have to expand or create new testbenches to exercise those statements with zero count.

3.4 Synthesis

After verifying the simulation results of the design, the next step is to synthesize the design. Synthesis is the process of converting and mapping the RTL verilog code into logic gates based on a standard cell library.

The process of synthesis requires three separate inputs:

1. standard cell library

2. design constraints

3. RTL design code

Synthesis can be categorized into pre-layout synthesis and post-layout synthesis. Pre-layout synthesis is synthesis on the RTL code using estimation on the interconnects between gates. Pre-layout synthesis uses wireload models which are statistical models of estimation on interconnects. Post-layout synthesis is an incremental synthesis process that is performed after layout. The interconnects between gates are accurately extracted after layout and back annotated for post-layout synthesis.

3.4.1   Standard Cell Library

A standard cell library is a library that consists of many different types of logic gates (and, or, not, nand, nor, xor, flip-flop, latch, and-nor, or-nand, and many others) with different types of sizing. A standard cell library normally consists of the following:

- basic logic gates
    i. and_a, and_b, and_c
   ii. or_a, or_b, or_c
  iii. nand_a, nand_b, nand_c
   iv. nor_a, nor_b, nor_c
    v. not_a, not_b, not_c, not_d, not_e, ... not_j
   vi. xor_a, xor_b, xor_c
- complex logic gates
  vii. and_nor_a, and_nor_b, and_nor_c (refer to Figure 3.21)
 viii. or_nand_a, or_nand_b, or_nand_c (refer to Figure 3.22)
   ix. or_or_nand_a, or_or_nand_b, or_or_nand_c (Refer to Figure 3.23)
    x. and_and_nor_a, and_and_nor_b, and_and_nor_c (refer to Figure 3.24)
- Registers (refer to Figure 3.25)
   xi. d_flop_a, d_flop_b, d_flop_c
  xii. reset_flop_a, reset_flop_b, reset_flop_c
 xiii. set_flop_a, set_flop_b, set_flop_c
  xiv. set_reset_flop_a, set_reset_flop_b, set_reset_flop_c

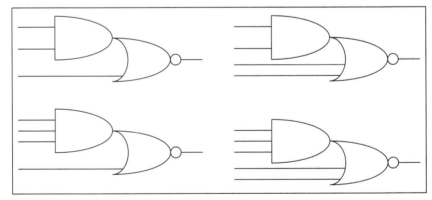

Figure 3.21   Diagram showing different types of and_nor gates.

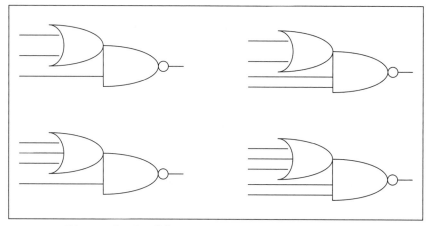

Figure 3.22   Diagram showing different types of or_nand gates.

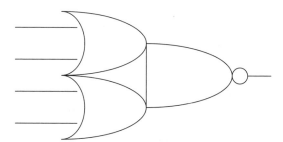

Figure 3.23   Diagram showing an or_or_nand gate.

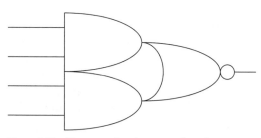

Figure 3.24   Diagram showing an and_and_nor gate.

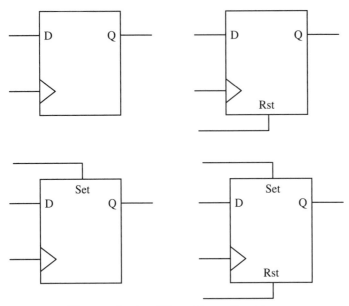

Figure 3.25 Diagram showing different types of registers.

- Latches (refer to Figure 3.26)
    - xv.   d_latch_a, d_latch_b, d_latch_c
    - xvi.   reset_latch_a, reset_latch_b, reset_latch_c
    - xvii. set_latch_a, set_latch_b, set_latch_c
    - xviii. set_reset_latch_a, set_reset_latch_b, set_reset_latch_c

All the gates in the standard cell library end with an alphabet. The alphabet represents the size of the gate. A larger alphabet represents a larger size gate which has larger drive strength.

The standard cell library is an important requirement during synthesis, as the RTL code is mapped to the logic gates of the standard cell library.

Size of a standard cell library varies greatly between different designs. Typically a standard cell library has at least 50 types of gates to several hundred types of gates. A larger standard cell library can have better synthesis optimization compared to a smaller standard cell library. However, a large standard cell library is difficult to create and maintain. The size of a standard cell library is dependent on the type of application for which the design is targeted. Designs targeted for high speed performance commonly have a large standard cell library that consists of hundreds of logic gates while designs that do not require high speed performance commonly have a smaller standard cell library.

Figure 3.27 shows a flow to create a standard cell library.

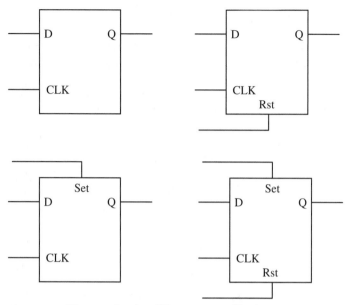

Figure 3.26 Diagram showing different types of latches.

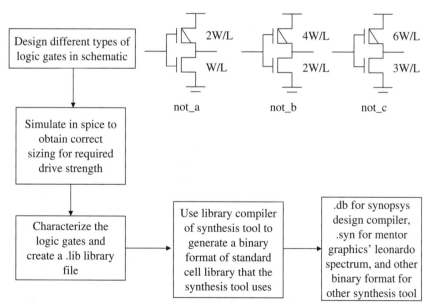

Figure 3.27 Diagram showing flow for creating a standard cell library.

In Figure 3.27, characterization of the standard cell logic gates creates a .lib (Synopsys Liberty format) file which is readable as text. This file contains all the information about the different logic gates, its input capacitance, its area, fan-out information, and timing information on each pin of the logic gate.

There are many ASIC synthesis tools available in the market, for example, Synopsys's Design Compiler, Mentor Graphics' Leonardo Spectrum, Cadence's Ambit, and many others. Each of these synthesis tools uses its own binary format for the standard cell library. An example is Mentor Graphics' Leonardo Spectrum which uses .syn format for its standard cell library. The synthesis tool provides for a Library Compiler that can compile the .lib text file into a corresponding binary format that the synthesis tool can use for synthesis process.

3.4.2 Design Constraints

During the process of synthesis, the synthesis tool reads in the RTL code and maps it into logic gates based on the standard cell library. Apart from requiring an RTL code and standard cell library, the process of synthesis also requires design constraints.

Design constraints specify requirements of the synthesized circuit, for example:

1. What is the clock frequency?
2. Should synthesis focus on synthesizing a circuit for performance or should it synthesize for area optimization?
3. What is the allowed fan-out for the logic gates?
4. Are there any multicycle paths?
    i. Multicycle paths are paths in a design that require more than one clock cycle (Refer to Figure 3.28).
    ii. Multicycle paths must be specified during synthesis to "inform" the synthesis tool that a particular path requires more than one clock cycle. Otherwise, the synthesis tool may spend a lot of its computational resource to optimize that path when there is no necessity for it since the path requires more than one clock cycle.
5. Are there any false paths?
    iii. False paths are paths that are asynchronous in nature and can occur at any given time, irrespective of clock reference.
    iv. False paths must be specified to allow the synthesis tool to understand which paths are false. An example of a false path is reset path.

Most synthesis tools also have additional commands that can be used as design constraints to allow for an efficient synthesis either in terms

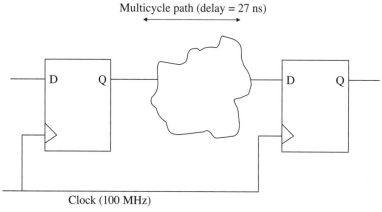

Multicycle path (delay = 27 ns)

Clock (100 MHz)

Figure 3.28 Diagram showing a multicycle path.

of performance optimization or area optimization. This is commonly referred to as synthesis tweaks. Several methods of performing synthesis tweaks are described in Section 3.4.3.

Appendix B shows the synthesis results of the VLIW microprocessor with timing and area report. The output of synthesis is a structural gate level netlist which is passed to layout. Appendix B shows the structural gate level netlist of the VLIW microprocessor generated from synthesis.

3.4.3   Synthesis Tweaks

When an RTL code is synthesized, it is common that the synthesized circuit is unable to meet requirements, either in terms of performance or area utilization. The designer will need to perform synthesis tweaks to squeeze the synthesis tool to obtain better and improved synthesis results. There are several ways to tweak the synthesis process:

1. *Register/logic balancing.* This method balances the amount of logic from one path to another path, thereby allowing better performance for all paths. This method is described in detail in Section 3.2.1.1

2. Decode *early arriving signals compared to late arriving signals.* Some signals may arrive at a given node earlier or later compared to others. For circuits that have signals that are late, the decoding logic can be moved to decode early arriving signals, thereby reducing the overall path delay. Figure 3.29 shows a logic circuit that have a late arriving signal.

   Referring to Figure 3.29, the worst path delay is

   $$\text{Delay} = 7.2 \text{ ns} + 1.5 \text{ ns} + 12 \text{ ns} = 20.7 \text{ ns}$$

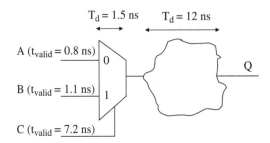

$T_d = 1.5$ ns $T_d = 12$ ns

A ($t_{valid} = 0.8$ ns)   0

B ($t_{valid} = 1.1$ ns)   1

Q

C ($t_{valid} = 7.2$ ns)

Figure 3.29 Diagram showing signal C as late arriving signal.

In order to improve this circuit, the decoding logic can be brought forward before the multiplexer to allow decoding of the early signals A and B, as shown in Figure 3.30.

Referring to Figure 3.30, the worst path delay is

$$\text{Delay} = 1.1 \text{ ns} + 12 \text{ ns} + 1.5 \text{ ns} = 14.6 \text{ ns}$$

This method is referred to as logic duplication, whereby the combinational logic after the multiplexer is duplicated and brought before the multiplexer allowing the two earlier arriving signals (A and B) to be decoded before late arriving signal C is valid.

3. *Using multiplex encoding and priority encoding.* Figure 3.31 shows an example of multiplex encoding and priority encoding. Priority encoding is normally used when any of the signals are late arriving. In the example shown in Figure 3.31, signal D is the late arriving signal. Multiplex encoding is normally used when all the signals are valid at the same time.

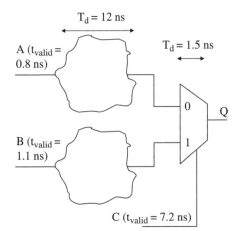

$T_d = 12$ ns

A ($t_{valid} = 0.8$ ns)

$T_d = 1.5$ ns

0

Q

B ($t_{valid} = 1.1$ ns)

1

C ($t_{valid} = 7.2$ ns)

Figure 3.30 Diagram showing decoding of early arriving signals A and B.

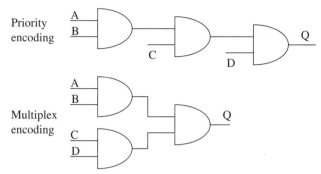

Figure 3.31   Diagram showing priority encoding and multiplex encoding.

4. *All synthesis tools have their own set of synthesis commands that allow the designer to squeeze synthesis to obtain better synthesis results.* Some tools allow for weightage on a net and some tools allow for critical level set on a net, but both allow the synthesis tool to focus its optimization on those nets.

3.5   Formal Verification

When synthesis is completed, designers will perform a formal verification on the synthesized netlist. This process compares the synthesized netlist and the RTL code to ensure that the synthesized circuit matches the RTL code, using equivalence checking techniques.

If a mismatch occurs during formal verification, the designer must look into the synthesis process as well as the RTL code to debug the source of the mismatch. The mismatch may be caused by nonsynthesizable verilog code in the design, or by errors introduced during the synthesis process. Figure 3.32 shows a flow used for formal verification.

There are many formal verification tools used in the industry, including Incisive, Formality, and FormalPro.

3.6   Pre-layout Static Timing Analysis

Static timing analysis is the process of timing verification that verifies a design for setup time violation and hold time violation.

Setup time violation occurs when a path takes longer than the required time. If a path has setup time violation, that path is too slow compared to the required timing. To fix a setup time violation, the path must be optimized for faster timing. Figure 3.33 shows a path that has a setup time violation.

Referring to Figure 3.33, the rising clock edge flip-flop has a setup time requirement of 1 ns. In order for the flip-flop to capture the data at

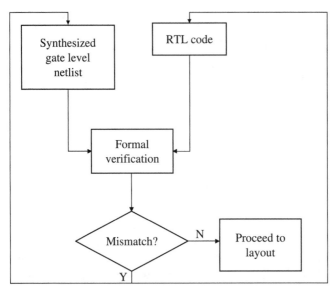

Figure 3.32   Diagram showing flow for formal verification.

input D during the rising edge of clock, the data at input D must be valid before the setup time requirement of the flip-flop. In this case, the signal at netB must be valid at least 1 ns before the rising edge of clock. Figure 3.34 shows a timing diagram of signal netB with a setup time requirement of the flip-flop.

Referring to Figure 3.34, the signal netB is valid at time t_x before the rising edge of clock. Signal netB meets the setup time requirement if $t_x > t_{setup}$. Based on this requirement, the circuit shown in Figure 3.33 has a setup time violation because the signal netB is valid after the rising edge of clock. In order for the circuit of Figure 3.33 to meet setup

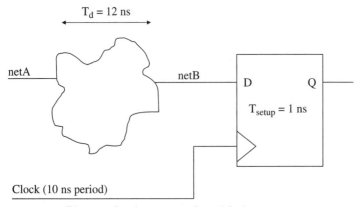

Figure 3.33   Diagram showing a setup time violation.

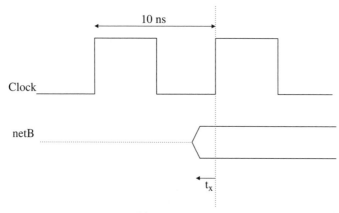

Figure 3.34 Diagram showing timing of netB with a setup time requirement.

time, the delay T_d of the logic block must be reduced to less than (clock period $-$ t_{setup}). If the circuit of Figure 3.33 is optimized to a delay of $T_d = 9$ ns or lower, the circuit is able to meet the setup time requirement.

Hold time violation is a violation that occurs when a path does not hold its signal valid for a minimum amount of time. If a path has hold time violation, that path is "too fast" compared to the required timing. To fix a hold time violation, the path must be slowed down by inserting buffers. Figure 3.35 shows a path that has a hold time violation.

Referring to Figure 3.35, hold time violation occurs when netB holds its value valid for a time less than the specified hold time of the flip-flop ($t_x < T_{hold}$).

During pre-layout static timing analysis, the synthesized gate level netlist and estimated interconnect delay from synthesis is used to build a pre-layout static timing model. The timing of each path within the synthesized circuit is calculated and a timing report is generated. Paths that have setup violation and hold violation are reported. This information is used for synthesis tweaks to reoptimize those failing paths. Paths that

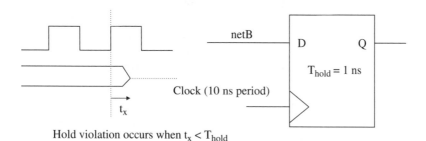

Hold violation occurs when $t_x < T_{hold}$

Figure 3.35 Diagram showing a hold time violation.

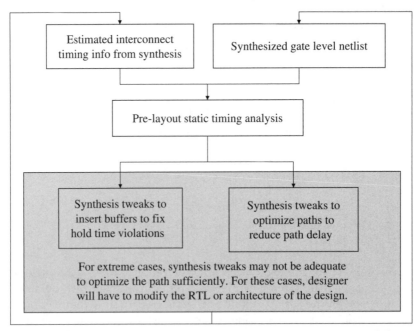

Figure 3.36　Diagram showing flow for fixing pre-layout static timing analysis failures.

have setup violation are tweaked for optimization to shorten the delay of the path while paths that have hold violation will have buffers inserted into them. For a better understanding of synthesis tweaks that can be used for optimization during synthesis, please refer to *VHDL Coding and Logic Synthesis With Synopsys* by Weng Fook Lee (Academic Press).

Since this process is pre-layout and the interconnect delays are estimated, not all the failing paths need to be fixed. Most designers will fix only those paths with timing failure greater than 10% of the setup and hold time requirement. Those failing within 10% of the setup and hold requirement are usually ignored.

For some extreme cases, synthesis tweaks for optimization may not be adequate to obtain sufficient timing to get the failing path to pass. For example, if a failing path needs to be optimized and reduce its delay by 30% or greater, this failing path will likely fail after synthesis tweaks. For such cases, the designer will have to modify the RTL code or the architecture of the design. Figure 3.36 shows the flow used for fixing setup and hold time violations in pre-layout static timing analysis.

3.7　Layout

Upon completion of pre-layout synthesis as mentioned in Section 3.4, the synthesized gate level netlist is passed to layout. In the layout process, the synthesized circuit is implemented using physical fabrication layers.

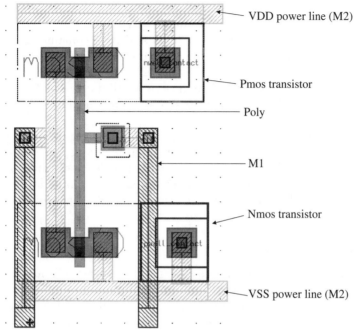

VDD power line (M2)

Pmos transistor

Poly

M1

Nmos transistor

VSS power line (M2)

Figure 3.37 Diagram showing physical layout of an inverter (layout is generated using Mentor Graphics IC Station SDL).

During this process, the layout designer uses different layers (poly, metal, n+, p+, and others) to form transistors, logic gates, resistors, and capacitors. Figure 3.37 shows an example of layout of an inverter (layout design in Mentor Graphics' IC Station).

There are three methods of performing layout:

1. manual/custom layout

2. semi-custom/auto layout

3. auto place and route

Each of the methods requires a different amount of engineering resource and each has its own set of layout issues that must be addressed carefully.

3.7.1 Manual/Custom Layout

Manual/custom layout as the name implies is based on layout performed manually by a layout designer. Manual/custom layout is tedious and time consuming as all transistors, logic gates, capacitors, and resistors are drawn manually. The size of each transistor (W/L) is manually measured and drawn in layout. Examples of some manual layout tools are Mentor Graphics' IC Station and Cadence's Virtuoso.

Although manual/custom layout is tedious and time consuming, it creates layouts that are smaller and therefore creates silicon dies that are compact and lower in cost. Due to its smaller die size, custom layout also provides for better design performance. A small die translates to shorter interconnects, which in turn translates to smaller interconnect RC parasitic.

Manual/custom layout is commonly used, especially for analog circuits or sensitive circuits such as sense amplifiers, charge pumps, cascades, IO buffers, phase locked loops (PLL), and others. It is also performed on digital circuits that require high speed performance, as manual layout allows for much flexibility in creating the shortest possible path for critical signals.

3.7.2 Semi-custom/Auto Layout

This layout process, as the name implies, is partially automated and partially manual. It is also referred to as schematic driven layout. In this layout process, the transistor or gate level layout is automatically generated by the layout tool. An example of such a tool is Mentor Graphics' IC Station SDL.

In this process, the layout tool reads the schematic of a design and allows the designer to point and click on a particular transistor or logic gate in the schematic. The layout tool will generate the layout of the mentioned transistor or logic gate based on the W/L parameters indicated in the schematic. The layout designer can manually place the generated layout anywhere in the floorplan. This process is repeated for the other transistors/gates/resistors/capacitors and manually placed.

Upon completion of transistor or logic gate placement, the layout tool can auto-route all the components together. If the layout designer is not satisfied with the automated routing interconnects, those interconnects can be deleted and manually routed.

The process of semi-custom/auto layout is a more efficient means of layout than manual/custom layout. However, the die area obtained from this layout process is larger than manual/custom layout, although it does allow the designer to design the layout in a much shorter timeframe.

3.7.3 Auto Place and Route

Auto place and route, more commonly known as APR (some designers refer to this process as BPR—block place and route), is a very different layout process. In the manual/custom layout and semi custom/auto layout, there is manual intervention from the layout designer. In APR, the layout is performed automatically by an APR tool. An example of an APR tool is Synopsys's Astro and Cadence's Silicon Ensemble.

In this method, the synthesized gate level netlist from synthesis is read into the APR tool together with a standard cell library. The standard cell library used by the APR tool is similar to the standard cell library used during synthesis. In APR, the standard cell library consists of the layout of all the different types of logic gates defined in the standard cell library for synthesis.

During APR, the tool maps the synthesized gate level netlist to the corresponding layout of each cell as specified in the APR's standard cell library. Each cell is placed automatically and the APR tool will route them automatically.

APR is a very efficient method of layout. It requires little time to complete a layout of a multimillion gate design compared to the other two. However, there is a disadvantage to using APR. Layout generated from APR tools is larger compared to the other two methods. It is also slower in performance due to the bigger interconnect RC parasitic as the interconnect routes are longer.

When considering these three methods of layout, the chosen method is largely dependent on the type of chip being designed. ASIC chips which are multimillion gate size are commonly APR as the time required to do manual/custom layout or semi-custom/auto layout is too long.

On the other hand, IC designs such as SOC or mixed signal IC chips commonly use a combination of all three layout methods. The analog circuits, sensitive circuits, and high speed digital circuits within SOC are manual/custom layout or semi-custom/auto layout, while the remaining digital circuits which do not require performance are APR.

3.8   DRC / LVS

When layout is completed, the layout must be verified for a set of design rules specified by the fab. For example, if the VLIW microprocessor is to be fab by Silterra's 0.35 micron process, the layout of the VLIW microprocessor must be verified by a set of design rules specified by Silterra for its 0.35 micron process technology. The design rules specify rules of fabrication that must be met prior to fabrication. Some examples of design rules are as follows:

- minimum active area width
- minimum active area spacing
- minimum N channel body implant spacing
- minimum poly width
- minimum poly spacing

- minimum N+ implant spacing
- minimum contact spacing
- minimum contact to gate spacing
- minimum metal width
- minimum metal spacing

During the design rule check (DRC) process, a check for any violations to the fab's specified rules is performed. If any violations occur, the layout designer must fix them in layout. DRC is performed again to verify the fixes. This is repeated until the design does not have any DRC violations.

Once DRC is verified clean, the layout designer will perform layout versus schematic (LVS). In LVS, layout is verified to match the schematic (the schematic can be a custom designed schematic or a synthesized schematic). If any violations occur (layout does not match schematic), the layout designer will have to fix these violations. When DRC and LVS are both verified clean, the design proceeds to RC extraction.

3.9   RC Extraction

As described in Section 3.4, during pre-layout synthesis the interconnect delays are estimated based on statistical wireload model. Therefore, the timing information obtained is inaccurate and estimated. When layout is completed with DRC and LVS clean, the accurate delay for each interconnect is extracted and calculated in RC extraction. The extracted information can be in the form of sdf (standard delay format), dspf, or spef format. This extracted delay is used for post-layout logic verification and post-layout performance verification.

3.10   Post-layout Logic Verification

This process is referred to as gate level simulation. It involves resimulation of the design using the gate level netlist, extracted gate level delay, and interconnect delay from RC extraction. This step allows for an accurate simulation of the gate level functionality with accurate delay of each gate and net. Any functionality failure during this process may be caused by timing issues such as race conditions or glitches. Any failures caught must be fixed using synthesis tweaks for incremental synthesis, layout tweaks for incremental layout improvements, or in some extreme conditions rewriting the RTL code. Figure 3.38 shows the flow for post-layout logic verification.

Post-layout logic verification uses large amounts of computation power and requires long simulation time due to the large amount

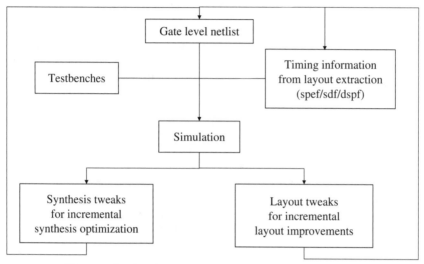

Figure 3.38   Diagram showing design flow for post-layout logic verification.

of delay information for gate and interconnect obtained from RC extraction.

3.11   Post-layout Performance Verification

This process is performed in parallel with post-layout logic verification. Its main objective is to catch any timing problems such as setup and hold violations in the design. This process is similar to the pre-layout static timing analysis except that the gate delays and interconnect delays are now accurate.

During this process, any setup and hold violations caught are fixed using either synthesis tweaks for synthesis incremental optimizations or incremental layout improvements. Its design flow shown in Figure 3.39 is similar to the pre-layout static timing analysis.

3.12   Tapeout

When all the violations and failures of post-layout logic verification and post-layout performance verification have been fixed and verified, the design is checked to ensure DRC/LVS is clean. When this is achieved, the design is ready for tapeout.

During this process, a GDSII file is generated from layout and this file is passed to mask making for fabrication of the ASIC device.

There are occasions when a design is tapeout without all violations being fixed. This occurs when the violations are taught to be false violations which never happens in a real life use of the design.

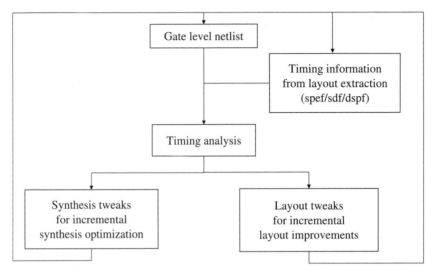

Figure 3.39 Diagram showing design flow for post-layout performance verification.

Such violations are waived by the project managers before proceeding to tapeout.

3.13   Linking Front end and Back end

The front end portion of the design flow is categorized as those steps that involve design, simulation, and synthesis. The back end portion includes the layout, physical verification, and physical extraction for back annotation.

It is common in today's complex devices that most designs that pass logic verification and timing analysis in the pre-layout phase of the flow will fail when layout parasitic is back annotated into the design. Synthesis tweaks are performed to fix the failures. For extreme cases, synthesis tweaks alone are not adequate and may require some RTL recoding or even architectural changes. Once the fixes are made, the whole flow is executed again. This is repeated until the design converges and post-layout timing analysis and functional verification passes. Only then can tapeout occur.

To tapeout a design, there are several iterations of the design flow. Fewer iterations are desired to achieve convergence of a design and tapeout. Several methods are used by designers to minimize the number of iterations:

1. *Floorplanning.* In floorplanning, different groups of circuits are categorized into certain portions of the chip. The objective is to achieve as few "long" interconnects as possible during layout of the fullchip. Different circuits that have many interconnects between them are

placed close to each other in order to avoid long interconnect lines between them. Long interconnect lines result in large parasitic and increases the routing area of the chip. Having a good floorplan is important to achieve a small layout and minimize the effects of parasitic. Floorplanning is not limited to the layout phase but is also used prior to synthesis. A good floorplan will group logic with similar features and functionality into groups, thereby allowing for more optimal logic sharing during synthesis. It also reduces the amount of nets between different groups of logic.

2. *Forward annotation.* After synthesis, the synthesized gate level netlist is passed to layout. In forward annotation, timing information from synthesis is forward annotated to the layout phase. This provides information to the layout tool on expected timing between gates and nets. Some synthesis tools have built-in layout placement algorithms that allow the synthesis tool to predict the layout placement, thus allowing better optimization during synthesis.

3. *Layout of clock tree.* Clock is the most important signal in fullchip. The clock signal is routed to all the clock ports of every flip-flop and latch. Different clock routing to different flip-flop/latches will have different clock skew. If the difference in clock skew between the flip-flops/latches is large, the design may fail. Figure 3.40 shows a fullchip design with several different flip-flops placed across different locations on the fullchip. In this figure, flip-flops A, B, and C are located near to each other, while flip-flop D is further away.

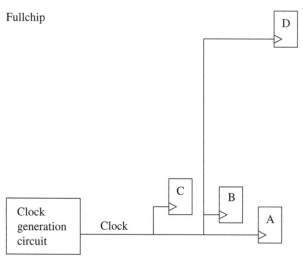

Figure 3.40 Diagram showing clock routing for different flip-flops.

Each flip-flop has a different length of clock routing connected to it. The parasitic on each of the clock routing is different. When the clock generation circuit generates a clock pulse, each of the flip-flops will see a rising edge of clock signal at different times. The larger the parasitic on the clock routing, the larger will be its clock skew. Flip-flops A, B, and C are relatively near to each other, thus the clock skew for these three flip-flops are negligible. However, because flip-flop D is placed far away, its clock skew is larger. Figure 3.41 shows the clock skew at the different flip-flops.

Flip-flops A, B, and C will flop the data at their output at approximately the same time. Flip-flop D will only flop the data at a later time, as the clock takes a longer time to reach flip-flop D. The output of flip-flops A, B, C, and D are valid at different times although they are clocked by the same clock signal. This difference may cause the design to fail.

To avoid this problem, placement of flip-flops in layout have the highest priority with the clock network being the first to be routed. During the process of APR, the first step is to layout the clock tree (some designers refer to this as clock tree synthesis). During clock tree layout, if a certain clock branch on the clock network has its clock skew drifting away from the specified clock skew (due to heavy parasitic), the APR tool will automatically insert buffers to the clock branch. For manual/custom layout and semi-custom/auto layout, the insertion of clock buffers are manually performed. Whether the clock buffers are

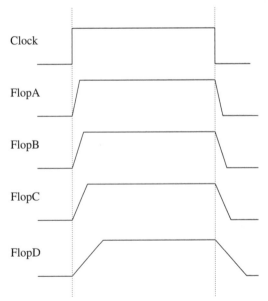

Figure 3.41   Diagram showing clock skew of different flip-flops.

inserted manually or automatically, the objective is the same—to achieve a clock network with a clock skew within specified target.

4. *Layout of critical nets.* After layout of the clock tree, the next step is to layout the critical nets of a design. Prior to layout, designers will create a list of critical signals which are given high priority for routing. The objective is to obtain layout routing with as minimal parasitic as possible on these critical signals.

5. *Back annotation.* As described in Section 3.9, once layout is completed and DRC/LVS is clean, RC extraction is performed to extract the parasitic RC. This information is then back annotated to the design to allow the design to resimulate with accurate information on gate and interconnect delays.

3.14   Power Consumption

Power consumption is an important part of design. Large adoption of handheld devices utilizing batteries has created a need for IC chips that are energy efficient. Designing devices that have low power consumption allows the devices to have longer usage time per single charge of battery.

This need for low power consumption has lead design to change its VCC or power voltage from 5V to 3V and is currently at 1.8V. There are new devices operating at 1.5V and lower. Using a lower VCC voltage would translate to less power consumption.

To address this requirement, there are several ways that designers use for their design:

1. Using a clock generation circuit that is able to vary its clock speed to a lower frequency when the device is not plugged in to a power socket. By moving to a lower clock frequency, the power consumption reduces.

2. Use clock gating to disable clock to localized portions of circuit in a design.

   In this method, a custom designed AND gate is used to block the clock using an enable signal. Figure 3.42 shows an example of a gated clock.

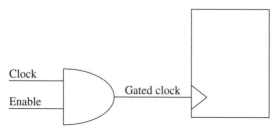

Figure 3.42   Diagram showing an AND Gate for gated clock.

In Figure 3.42, the AND gate is custom built specifically for use for clock gating. A normal AND gate from a standard cell library is not used because special requirements such as clock skew control, clock fanout, and clock driving strength are needed for the AND gate used for gated clock. Some designers choose to use a custom built OR gate for clock gating instead of a custom built AND gate.

3.15   ASIC Design Testability

ASIC chips commonly consist of multi-million gates due to the many features and functionality cramped into a single ASIC chip. With this complexity, testing the ASIC chip becomes an issue due to the large amount of logic and functionality.

During design phase, designers have to take into consideration DFT (design for testability) to ensure that the chip can be easily tested for defects. There are several ways in which designers approach testability issues:

1. *BIST (built-in self-test).* In this method, designers design built-in special circuits that function to execute tests to determine if certain features on the ASIC chip are functional. Ideally, the BIST circuit can execute tests to check for all the features and functionality of the ASIC chip. However, checking all features and functionalities are not practical as the BIST circuits will be too large. Therefore, BIST circuits are designed to self-test certain critical functions or features without the use of complex and expensive testers. Use of BIST increases die size and therefore increases die cost. However, it reduces test cost as some of the functionalities and features are self-tested by the ASIC chip.

2. *JTAG (boundary scan and scan chain).* JTAG is a method developed by a group called Joint Test Action Group. The group developed a method of testing referred to as boundary scan. In this method, the registers of the design are replaced by scan registers. Figure 3.43 shows the difference between a register and a scan register. A scan register is a data register with a multiplexer to multiplex either data

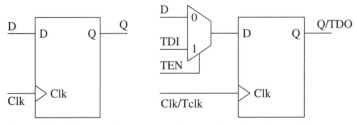

Figure 3.43   Diagram showing a register replaced by scan register.

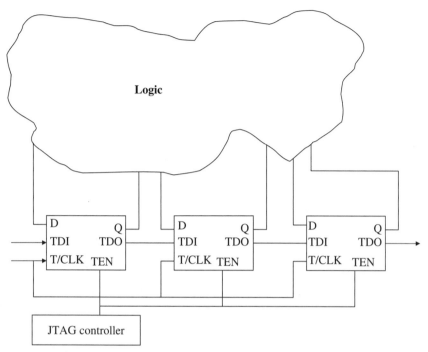

Figure 3.44   Diagram showing a boundary scan chain with JTAG state controller.

or TDI (test data in). The clock of the scan register is controlled by either the clock of the ASIC chip during normal operations or by TClk (test clock) during testing. The signal TEN (test enable) determines if the scan register is to function as a "normal" register or a scan register by controlling the select path of the multiplexer.

The scan registers are stitched together to form a boundary scan chain through the scan path and output pin TDO. When multiple IC/ASIC chips are placed on a board, the connection of these pins allows testing of external interconnection between different IC/ASIC chips, as well as testing of internal logic within the IC/ASIC chip.

Each IC/ASIC chip that uses this method has a 16-state controller (TAP controller) that controls the propagation of these test signals. Figure 3.44 shows different scan registers stitched to form a boundary scan chain.

In Figure 3.44, test data can be inserted into the design using TDI, serially moved from one scan register to the next through the scan chain before propagating to the output pin TDO.

3. *Custom DFT.* Some designers design built-in custom testability circuits that are used for specialized in-house testing.

FPGA Implementation

Figure 2.1 shows the design methodology flow for implementation of an ASIC design. Sections 3.4 to 3.12 in Chapter 3 describe the steps involved in synthesis to tapeout of an ASIC design. There is another alternative available to designers that do not wish to proceed with the ASIC path. Field-programmable gate array (FPGA) implementation is a different method of transforming the design into an IC device.

There are several reasons why some designers prefer FPGA implementation to ASIC implementation. Each has its own set of advantages and disadvantages and provides for very different cost structures. Section 4.1 describes the differences between FPGA and ASIC implementation.

4.1 FPGA Versus ASIC

An FPGA is an IC chip that allows designers to "download" their digital circuit into the IC chip and allows the IC chip to function as described by the downloaded digital circuit. A designer can design many different digital circuits and program these digital circuits into the FPGA. Among the many FPGAs that are available in the market are Altera, Xilinx, Actel and Atmel. FPGAs are programmable and allow for ease of configuration of digital circuits.

Conversely, an ASIC is an IC chip that specifically caters only to a digital circuit for which the ASIC chip was designed. Table 4.1 shows a detailed description of the advantages and disadvantages between each.

TABLE 4.1   Advantages and Disadvantages of FPGA and ASIC

FPGA	ASIC
FPGA can be "reprogrammed" to function as different digital circuits.	ASIC chip once fabricated can only be used as the circuit for which it was designed.
Per unit cost of an FPGA device is high compared to an ASIC device.	Per unit cost of an ASIC is lower than an FPGA device.
Overall cost of creating an IC chip for a particular design using FPGA is low. FPGAs are sold as an IC chip that the designer can use to program the digital circuit. Designing with FPGA does not require mask making, fabrication, creation of wafer, packaging, and package testing.	Overall cost of creating an IC chip for a particular design for ASIC is high (mask cost, fabrication cost, wafer saw, wafer sort, die attach, packaging, and package testing increases the overall cost of ASIC significantly compared to FPGA).
Time to completion of design is faster as there is no need for layout, DRC, or LVS.	Time to completion of design is long as layout, DRC, and LVS requires engineering resource and time.
FPGA consumes large amount of current and is not suitable for designs that require low power consumption.	ASIC consumes much less current and is ideally suited for designs that require low power consumption.
A digital circuit implemented in FPGA is slower. It is common to estimate that the same digital circuit implemented in FPGA can be 2× faster or more when implemented using the same process technology in ASIC.	A digital circuit implemented in ASIC is faster.
Suitable for use as a prototyping device as it allows for reprogrammability.	Not suitable for prototyping. A digital circuit implemented in ASIC cannot be modified without layout changes, DRC, LVS, mask remaking, and refabrication.
Suitable for low volume production as cost per unit is high.	Ideal for high volume production as cost per unit is low for high volume.

4.2   FPGA Design Methodology

The design methodology flow shown in Figure 2.1 is for an ASIC flow. An FPGA flow is similar to that of an ASIC; however, FPGA flow does not have any layout/DRC/LVS involved. The design is synthesized for an FPGA technology (for example, a design is synthesized to Altera's Cyclone2 technology or Xilinx's Spartan 2 technology) and placed and routed using that particular technology. The completed place and route data are then downloaded into the FPGA. Once the download is completed, the FPGA will function as an IC chip with the functionality and features of the design.

In Figure 4.1, after verifying the functionality of the design, synthesis is performed using the targeted FPGA technology. FPGA synthesis tools are different from ASIC synthesis tools. ASIC synthesis tools as described in Chapter 3, Section 3.4 use a standard cell library.

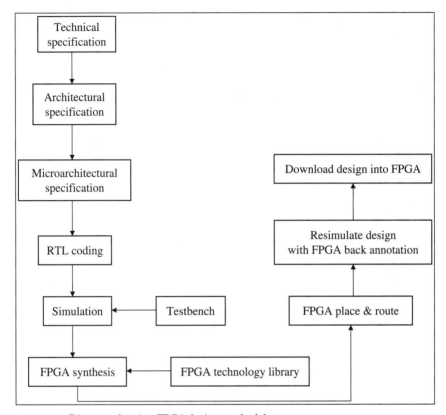

Figure 4.1 Diagram showing FPGA design methodology.

FPGA synthesis tools use an FPGA technology library which is provided by the FPGA supplier (Altera, Xilinx, Atmel, and others). Some examples of FPGA synthesis tools are Altera's Quartus II, Xilinx's ISE, Mentor Graphics' Precision, Synplicity's Synplify, and Synopsys's FPGA Express.

Once synthesis is completed, the design is placed and routed using the FPGA tools provided by the FPGA supplier (Altera's Quartus II, Xilinx's ISE). When completed, the designer can either choose to download the design onto the FPGA or alternatively recheck timing of the design by back annotating the delay information for resimulation.

4.3 Testing FPGA

FPGAs are widely used in the design industry due to their ease of usage and short design cycles. Not needing to do layout/DRC/LVS for a VLSI design tremendously reduces the required engineering resources.

Furthermore, most FPGAs today have built-in useful cores such as phase lock loop (PLL), clock management circuits, and PCI interfaces. These built-in cores allow designers to prototype a design with ease.

Once a design has been downloaded into an FPGA, the design must be tested for its functionality and features on the FPGA. As such, it is common for designers to use an FPGA development board. These boards consist of an FPGA and several other peripheral ICs, push buttons, an LCD display, an LED display, connectors to off-board function generators, and oscilloscopes, allowing for system level testing of the FPGA.

4.4 Structured ASIC

As described in Section 4.1, FPGAs have a high cost per unit. They are widely used to prototype a design and also for low volume design. However, if the volume increases, it does not make economical sense to continue to use FPGAs. The most cost-effective method for large volume production is ASIC, but ASIC requires a lot of engineering resource (layout/DRC/LVS) to transform the particular design into an ASIC design.

A more cost-effective and economical method for converting FPGA designs for medium-scale volume is to use structured ASIC. Structured

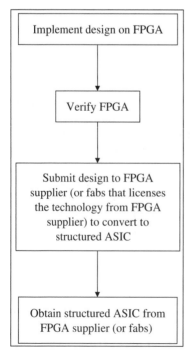

Figure 4.2 Diagram showing flow for conversion to structured ASIC.

ASIC is similar to FPGA and it also uses the technology of FPGA. Structured ASIC provides for an IC solution that has a lower cost per unit than FPGA but a higher cost per unit compared to ASIC, better power savings compared to FPGA but not as good as ASIC, and better speed performance compared to FPGA but not as good as ASIC. Structured ASIC is somewhere between FPGA and ASIC in terms of its capability. Figure 4.2 shows a general flow to convert a design from FPGA to structured ASIC. However, not all FPGA designs can be converted to structured ASIC. Each FPGA supplier has its selected target FPGA technology that allows for conversion to structured ASIC.

Appendix B shows the results for the FPGA synthesis of the VLIW microprocessor.

Testbenches and Simulation Results

Appendix A shows several of the testbenches used for verifying the VLIW microprocessor. A complete verification plan consists of many tests to fully verify each functionality and feature of the VLIW microprocessor. Example A.1 shows the verilog code for a testbench verifying the functionality of operation barrel shift left, subtract, multiply, and read.

Example A.1 Testbenches Verifying Barrel Shift Left, Subtract, Multiply, and Read

```
module vliw_top_tb();
reg clock, reset;
reg [191:0] data;
reg [63:0] word;
wire [63:0] readdatapipe1, readdatapipe2, readdatapipe3;
wire jump;
parameter halfperiod = 5;
parameter twocycle = 20;
parameter delay = 100;
// include the file that declares the parameter declaration for
//   register
// names and also instruction operations
`include "/project/VLIW/64bit/simulation/regname.v"
// clock generation
initial
begin
    clock = 0;
    forever #halfperiod clock = ~clock;
end
// pump in stimulus for vliw processor
initial
begin
    // do a reset
    data = 0;
```

```
setreserved;
setreset;
// word [58:55]opcode [53:50]src1 [48:45]src2 [43:40]dest op1
// word [38:35]opcode [33:30]src1 [28:25]src2 [23:20]dest op2
// word [18:15]opcode [13:10]src1 [8:5]src2   [3:0]dest    op3
// load all necessary values into r0 to r8
// load #123456789abcdef0, reg0 -> op1
word [58:55]   = 4'b0100; // load inst op1
word [53:50]   = reg0;    // src1 default to reg0 cause not used
word [48:45]   = reg0;    // src2 default to reg0 cause not used
word [43:40]   = reg0;
data [191:128] = 64'h123456789abcdef0; // data for op1
// load #1000000000000001, reg1 -> op2
word [38:35]   = 4'b0100; // load inst op2
word [33:30]   = reg0;    // src1 default to reg0 cause not used
word [28:25]   = reg0;    // src2 default to reg0 cause not used
word [23:20]   = reg1;
data [127:64]  = 64'h1000000000000001; // data for op2
// load #0111111111111110, reg2 -> op3
word [18:15]   = 4'b0100; // load inst op3
word [13:10]   = reg0;    // src1 default to reg0 cause not used
word [8:5]     = reg0;    // src2 default to reg0 cause not used
word [3:0]     = reg2;
data [63:0]    = 64'h0111111111111110;

// one clock delay
#halfperiod;
#halfperiod;

// load #abababababababab, reg3 -> op1
word [58:55]   = 4'b0100; // load inst op1
word [53:50]   = reg0;    // src1 default to reg0 cause not used
word [48:45]   = reg0;    // src2 default to reg0 cause not used
word [43:40]   = reg3;
data [191:128] = 64'habababababababab; // data for op1
// load #100000aaa19a8654, reg4 -> op2
word [38:35]   = 4'b0100; // load inst op2
word [33:30]   = reg0;    // src1 default to reg0 cause not used
word [28:25]   = reg0;    // src2 default to reg0 cause not used
word [23:20]   = reg4;
data [127:64]  = 64'h100000aaa19a8654; // data for op2
// load #01111111abc739ab, reg5 -> op3
word [18:15]   = 4'b0100; // load inst op3
word [13:10]   = reg0;    // src1 default to reg0 cause not used
word [8:5]     = reg0;    // src2 default to reg0 cause not used
word [3:0]     = reg5;
data [63:0]    = 64'h01111111abc739ab;

// one clock delay
#halfperiod;
#halfperiod;

// load #2121212123232323, reg6 -> op1
word [58:55]   = 4'b0100; // load inst op1
word [53:50]   = reg0;    // src1 default to reg0 cause not used
word [48:45]   = reg0;    // src2 default to reg0 cause not used
word [43:40]   = reg6;
data [191:128] = 64'h2121212123232323; // data for op1
// load #5a5a5a5aa5a5a5a5, reg7 -> op2
word [38:35]   = 4'b0100; // load inst op2
word [33:30]   = reg0;    // src1 default to reg0 cause not used
word [28:25]   = reg0;    // src2 default to reg0 cause not used
```

```
word [23:20]    = reg7;
data [127:64]   = 64'h5a5a5a5aa5a5a5a5; // data for op2
// load #9236104576530978, reg8 -> op3
word [18:15]    = 4'b0100; // load inst op3
word [13:10]    = reg0;     // src1 default to reg0 cause not used
word [8:5]      = reg0;     // src2 default to reg0 cause not used
word [3:0]      = reg8;
data [63:0]     = 64'h9236104576530978;

// one clock delay
#halfperiod;
#halfperiod;

// read r0 -> op1
word [58:55]    = 4'b0110; // read inst op1
word [53:50]    = reg0;     // src1 is reg0
word [48:45]    = reg0;     // src2 default to reg0 cause not used
word [43:40]    = reg0;     // dest default to reg0 cause not used
data [191:128]  = 0;        // not used
// read r1 -> op2
word [38:35]    = 4'b0110; // read inst op2
word [33:30]    = reg1;     // src1 is reg1
word [28:25]    = reg0;     // src2 default to reg0 cause not used
word [23:20]    = reg0;     // dest default to reg0 cause not used
data [127:64]   = 0;        // not used
// read reg2 -> op3
word [18:15]    = 4'b0110; // read inst op3
word [13:10]    = reg2;     // src1 is reg2
word [8:5]      = reg0;     // src2 default to reg0 cause not used
word [3:0]      = reg0;     // dest default to reg0 cause not used
data [63:0]     = 0;        // not used

// one clock delay
#halfperiod;
#halfperiod;

// barrel shift left r3, r4, r10 -> op1
word [58:55]    = 4'b1110; // barrel shift left inst op1
word [53:50]    = reg3;     // src1 is reg3
word [48:45]    = reg4;     // src2 is reg4
word [43:40]    = reg10;    // destination is reg10
data [191:128]  = 0;        // data is not used
// sub r0, r1, r11 -> op2
word [38:35]    = 4'b0010; // sub inst op2
word [33:30]    = reg0;     // src1 is reg0
word [28:25]    = reg1;     // src2 is reg1
word [23:20]    = reg11;    // destination is reg11
data [127:64]   = 0;        // data is not used
// mul r2, r1, r12 -> op3
word [18:15]    = 4'b0011; // mul inst op3
word [13:10]    = reg2;     // src1 is reg2
word [8:5]      = reg1;     // src2 is reg1
word [3:0]      = reg12;    // destination is reg12
data [63:0]     = 0;        // data is not used

// one clock delay
#halfperiod;
#halfperiod;

// read r3 -> op1
word [58:55]    = 4'b0110; // read inst op1
word [53:50]    = reg3;     // src1 is reg3
```

```
word [48:45]    = reg0;      // src2 default to reg0 cause not used
word [43:40]    = reg0;      // dest default to reg0 cause not used
data [191:128] = 0;          // not used
// read r4 -> op2
word [38:35]    = 4'b0110;   // read inst op2
word [33:30]    = reg4;      // src1 is reg4
word [28:25]    = reg0;      // src2 default to reg0 cause not used
word [23:20]    = reg0;      // dest default to reg0 cause not used
data [127:64]  = 0;          // not used
// read reg5 -> op3
word [18:15]    = 4'b0110;   // read inst op3
word [13:10]    = reg5;      // src1 is reg5
word [8:5]      = reg0;      // src2 default to reg0 cause not used
word [3:0]      = reg0;      // dest default to reg0 cause not used
data [63:0]     = 0;         // not used


// one clock delay
#halfperiod;
#halfperiod;


// read r10 -> op1
word [58:55]    = 4'b0110;   // read inst op1
word [53:50]    = reg10;     // src1 is reg10
word [48:45]    = reg0;      // src2 default to reg0 cause not used
word [43:40]    = reg0;      // dest default to reg0 cause not used
data [191:128] = 0;          // not used
// read r11 -> op2
word [38:35]    = 4'b0110;   // read inst op2
word [33:30]    = reg11;     // src1 is reg11
word [28:25]    = reg0;      // src2 default to reg0 cause not used
word [23:20]    = reg0;      // dest default to reg0 cause not used
data [127:64]  = 0;          // not used
// read reg12 -> op3
word [18:15]    = 4'b0110;   // read inst op3
word [13:10]    = reg12;     // src1 is reg12
word [8:5]      = reg0;      // src2 default to reg0 cause not used
word [3:0]      = reg0;      // dest default to reg0 cause not used
data [63:0]     = 0;         // not used


// one clock delay
#halfperiod;
#halfperiod;


// nop -> op1
word [58:55]    = 4'b0000;   // nop inst op1
word [53:50]    = reg0;      // src1 default to reg0 cause not used
word [48:45]    = reg0;      // src2 default to reg0 cause not used
word [43:40]    = reg0;      // dest default to reg0 cause not used
data [191:128] = 0;          // not used
// nop   -> op2
word [38:35]    = 4'b0000;   // nop inst op2
word [33:30]    = reg0;      // src1 default to reg0 cause not used
word [28:25]    = reg0;      // src2 default to reg0 cause not used
word [23:20]    = reg0;      // dest default to reg0 cause not used
data [127:64]  = 0;          // not used
// nop -> op3
word [18:15]    = 4'b0000;   // nop inst op3
word [13:10]    = reg0;      // src1 default to reg0 cause not used
word [8:5]      = reg0;      // src2 default to reg0 cause not used
word [3:0]      = reg0;      // dest default to reg0 cause not used
data [63:0]     = 0;         // not used
```

```
        // one clock delay
        #halfperiod;
        #halfperiod;

        #1000 $stop;
    end

    task setreserved;
    begin
        // all these bits in the vliw word are reserved and therefore not
        // used. they are meant for future expansion
        word [63:60] = 4'bxxxx;
        word [59] = 1'bx;
        word [54] = 1'bx;
        word [49] = 1'bx;
        word [44] = 1'bx;
        word [39] = 1'bx;
        word [34] = 1'bx;
        word [29] = 1'bx;
        word [24] = 1'bx;
        word [19] = 1'bx;
        word [14] = 1'bx;
        word [9] = 1'bx;
        word [4] = 1'bx;
    end
    endtask

    task setreset;
    begin
        // do a reset
        reset = 0;
        #twocycle;
        reset = 1;
        #twocycle;
        reset = 0;
        #twocycle;
    end
    endtask

    vliw_top vliw_top_inst (clock, reset, word, data, readdatapipe1,
readdatapipe2, readdatapipe3, readdatavalid, jump);
    endmodule
```

Figure A.1 shows the simulation results of Example A.1.

In this figure, the contents of register r10 are bababababababababa which are barrel shift left of contents of r3 (abababababababab) by lowest nibble of r4 (100000aaa19a8654). The contents of register r11 are 023456789abcdeef which are subtraction results of r1 (1000000000000001) from r0 (123456789abcdef0). The contents of register r12 are 0000000011111110 which are multiplication of r2 (0111111111111110) and r1 (1000000000000001). The upper 32 bits of register r12 are zero because for multiplication operation, the operands defined are only lower 32 bits (refer to Chapter 2, Section 2.1.2 on explanation of multiply operation).

Figure A.2 shows the simulation result of the read operation of Example A.1.

In this figure, when output port readdatavalid goes high, data at output port readatapipe1, readdatapipe2, readatapipe3 are

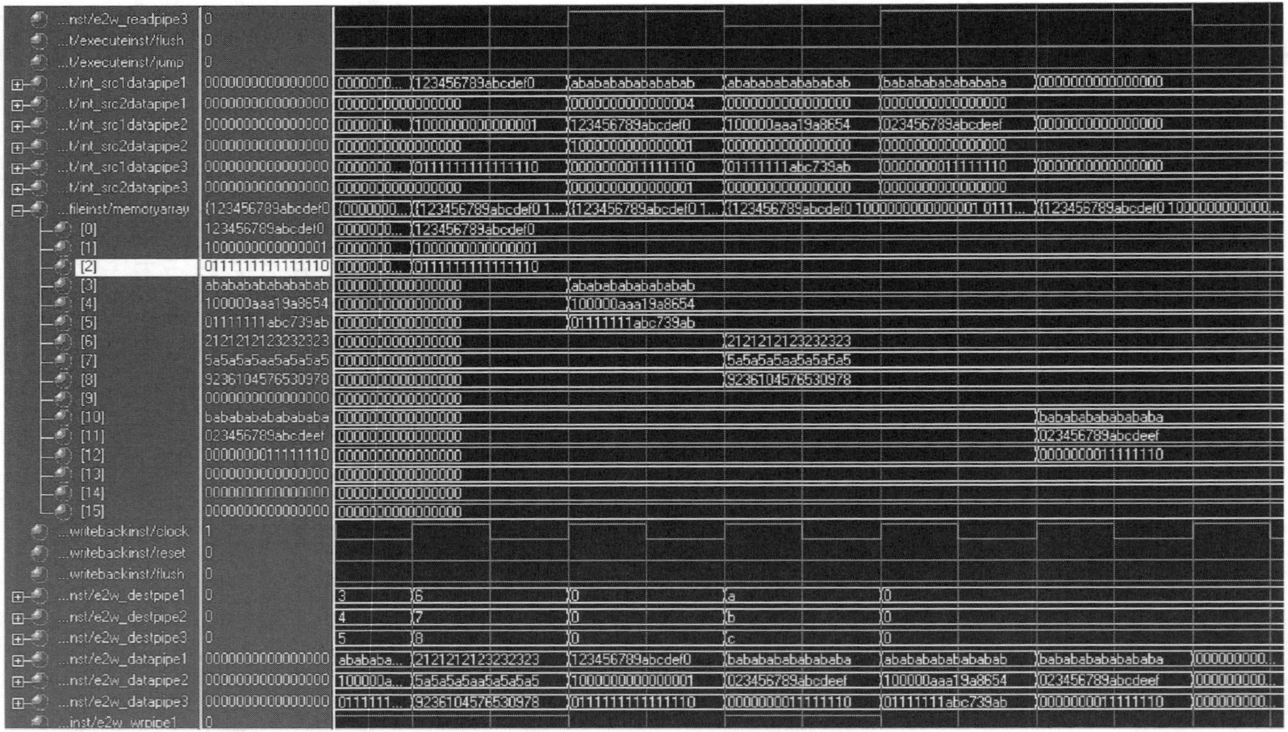

Figure A.1 Diagram showing simulation result of Example A.1 for barrel shift left, subtract, and multiply.

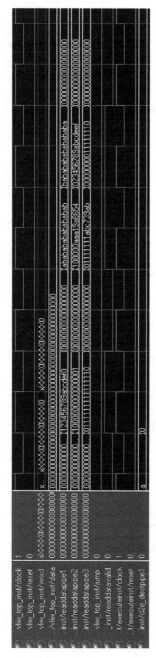

Figure A.2 Diagram showing read simulation result for Example A.1.

123456789abcdef0, 1000000000000001, and 0111111111111110
which are contents of register r0, r1, and r2. Subsequent reads are data
abababababababab, 100000aaa198654, 01111111abc739ab on
output port readdatapipe1, readdatapipe2, readdatapipe3 which
are contents of register r3, r4, and r5, followed by babababababababa,
023456789abcdeef and 0000000011111110 which are contents of
register r10, r11, and r12.

Example A.2 shows the testbench for simulating multiple register
bypass conditions between operation1, operation2 and operation3.

Example A.2 Testbench Verifying Multiple Register Bypass Condition between
Operation1, Operation2, and Operation3

```
module vliw_top_tb();

reg clock, reset;
reg [191:0] data;
reg [63:0] word;
wire [63:0] readdatapipe1, readdatapipe2, readdatapipe3;
wire jump;

parameter halfperiod = 5;
parameter twocycle = 20;
parameter delay = 100;

// include the file that declares the parameter declaration for register
// names and also instruction operations
`include "/project/VLIW/64bit/simulation/regname.v"

// clock generation
initial
begin
    clock = 0;
    forever #halfperiod clock = ~clock;
end

// pump in stimulus for vliw processor
initial
begin
    // do a reset
    data = 0;
    setreserved;
    setreset;
    // word [58:55]opcode [53:50]src1 [48:45]src2 [43:40]dest op1
    // word [38:35]opcode [33:30]src1 [28:25]src2 [23:20]dest op2
    // word [18:15]opcode [13:10]src1 [8:5]src2   [3:0]dest   op3

    // load all necessary values into r0 to r8
    // load #123456789abcdef0, reg0 -> op1
    word [58:55]    = 4'b0100; // load inst op1
    word [53:50]    = reg0;    // src1 default to reg0 cause not used
    word [48:45]    = reg0;    // src2 default to reg0 cause not used
    word [43:40]    = reg0;
    data [191:128] = 64'h123456789abcdef0; // data for op1
    // load #1000000000000001, reg1 -> op2
    word [38:35]    = 4'b0100; // load inst op2
    word [33:30]    = reg0;    // src1 default to reg0 cause not used
    word [28:25]    = reg0;    // src2 default to reg0 cause not used
    word [23:20]    = reg1;
```

```
data [127:64] = 64'h1000000000000001; // data for op2
// load #0111111111111110, reg2 -> op3
word [18:15]   = 4'b0100; // load inst op3
word [13:10]   = reg0;    // src1 default to reg0 cause not used
word [8:5]     = reg0;    // src2 default to reg0 cause not used
word [3:0]     = reg2;
data [63:0]    = 64'h0111111111111110;

// one clock delay
#halfperiod;
#halfperiod;

// load #ababababababababab, reg3 -> op1
word [58:55]   = 4'b0100; // load inst op1
word [53:50]   = reg0;    // src1 default to reg0 cause not used
word [48:45]   = reg0;    // src2 default to reg0 cause not used
word [43:40]   = reg3;
data [191:128] = 64'habababababababab; // data for op1
// load #100000aaa19a8654, reg4 -> op2
word [38:35]   = 4'b0100; // load inst op2
word [33:30]   = reg0;    // src1 default to reg0 cause not used
word [28:25]   = reg0;    // src2 default to reg0 cause not used
word [23:20]   = reg4;
data [127:64]  = 64'h100000aaa19a8654; // data for op2
// load #01111111abc739ab, reg5 -> op3
word [18:15]   = 4'b0100; // load inst op3
word [13:10]   = reg0;    // src1 default to reg0 cause not used
word [8:5]     = reg0;    // src2 default to reg0 cause not used
word [3:0]     = reg5;
data [63:0]    = 64'h01111111abc739ab;

// one clock delay
#halfperiod;
#halfperiod;

// load #2121212123232323, reg6 -> op1
word [58:55]   = 4'b0100; // load inst op1
word [53:50]   = reg0;    // src1 default to reg0 cause not used
word [48:45]   = reg0;    // src2 default to reg0 cause not used
word [43:40]   = reg6;
data [191:128] = 64'h2121212123232323; // data for op1
// load #5a5a5a5aa5a5a5a5, reg7 -> op2
word [38:35]   = 4'b0100; // load inst op2
word [33:30]   = reg0;    // src1 default to reg0 cause not used
word [28:25]   = reg0;    // src2 default to reg0 cause not used
word [23:20]   = reg7;
data [127:64]  = 64'h5a5a5a5aa5a5a5a5; // data for op2
// load #9236104576530978, reg8 -> op3
word [18:15]   = 4'b0100; // load inst op3
word [13:10]   = reg0;    // src1 default to reg0 cause not used
word [8:5]     = reg0;    // src2 default to reg0 cause not used
word [3:0]     = reg8;
data [63:0]    = 64'h9236104576530978;

// one clock delay
#halfperiod;
#halfperiod;

// read r0 -> op1
word [58:55]   = 4'b0110; // read inst op1
word [53:50]   = reg0;    // src1 is reg0
word [48:45]   = reg0;    // src2 default to reg0 cause not used
```

```
    word [43:40]   = reg0;     // dest default to reg0 cause not used
    data [191:128] = 0;        // not used
    // read r1 -> op2
    word [38:35]   = 4'b0110;  // read inst op2
    word [33:30]   = reg1;     // src1 is reg1
    word [28:25]   = reg0;     // src2 default to reg0 cause not used
    word [23:20]   = reg0;     // dest default to reg0 cause not used
    data [127:64]  = 0;        // not used
    // read reg2 -> op3
    word [18:15]   = 4'b0110;  // read inst op3
    word [13:10]   = reg2;     // src1 is reg2
    word [8:5]     = reg0;     // src2 default to reg0 cause not used
    word [3:0]     = reg0;     // dest default to reg0 cause not used
    data [63:0]    = 0;        // not used

    // one clock delay
    #halfperiod;
    #halfperiod;

    // add r4, r5, r10 -> op1
    word [58:55]   = 4'b0001;  // add inst op1
    word [53:50]   = reg4;     // src1 is reg4
    word [48:45]   = reg5;     // src2 is reg5
    word [43:40]   = reg10;    // destination is reg10
    data [191:128] = 0;        // data is not used
    // sub r3, r3, r11 -> op2
    word [38:35]   = 4'b0010;  // sub inst op2
    word [33:30]   = reg3;     // src1 is reg3
    word [28:25]   = reg3;     // src2 is reg3
    word [23:20]   = reg11;    // destination is reg11
    data [127:64]  = 0;        // data is not used
    // mul r2, r1, r12 -> op3
    word [18:15]   = 4'b0011;  // mul inst op3
    word [13:10]   = reg2;     // src1 is reg2
    word [8:5]     = reg1;     // src2 is reg1
    word [3:0]     = reg12;    // destination is reg12
    data [63:0]    = 0;        // data is not used

    // one clock delay
    #halfperiod;
    #halfperiod;

    // read r3 -> op1
    word [58:55]   = 4'b0110;  // read inst op1
    word [53:50]   = reg3;     // src1 is reg3
    word [48:45]   = reg0;     // src2 default to reg0 cause not used
    word [43:40]   = reg0;     // dest default to reg0 cause not used
    data [191:128] = 0;        // not used
    // read r4 -> op2
    word [38:35]   = 4'b0110;  // read inst op2
    word [33:30]   = reg4;     // src1 is reg4
    word [28:25]   = reg0;     // src2 default to reg0 cause not used
    word [23:20]   = reg0;     // dest default to reg0 cause not used
    data [127:64]  = 0;        // not used
    // read reg5 -> op3
    word [18:15]   = 4'b0110;  // read inst op3
    word [13:10]   = reg5;     // src1 is reg5
    word [8:5]     = reg0;     // src2 default to reg0 cause not used
    word [3:0]     = reg0;     // dest default to reg0 cause not used
    data [63:0]    = 0;        // not used

    // one clock delay
    #halfperiod;
    #halfperiod;
```

```
    // read r10 -> op1
    word [58:55]    = 4'b0110; // read inst op1
    word [53:50]    = reg10;   // src1 is reg10
    word [48:45]    = reg0;    // src2 default to reg0 cause not used
    word [43:40]    = reg0;    // dest default to reg0 cause not used
    data [191:128] = 0;        // not used
    // read r11 -> op2
    word [38:35]    = 4'b0110; // read inst op2
    word [33:30]    = reg11;   // src1 is reg11
    word [28:25]    = reg0;    // src2 default to reg0 cause not used
    word [23:20]    = reg0;    // dest default to reg0 cause not used
    data [127:64]  = 0;        // not used
    // read reg12 -> op3
    word [18:15]    = 4'b0110; // read inst op3
    word [13:10]    = reg12;   // src1 is reg12
    word [8:5]      = reg0;    // src2 default to reg0 cause not used
    word [3:0]      = reg0;    // dest default to reg0 cause not used
    data [63:0]    = 0;        // not used

    // one clock delay
    #halfperiod;
    #halfperiod;

    // nop -> op1
    word [58:55]    = 4'b0000; // nop inst op1
    word [53:50]    = reg0;    // src1 default to reg0 cause not used
    word [48:45]    = reg0;    // src2 default to reg0 cause not used
    word [43:40]    = reg0;    // dest default to reg0 cause not used
    data [191:128] = 0;        // not used
    // nop  -> op2
    word [38:35]    = 4'b0000; // nop inst op2
    word [33:30]    = reg0;    // src1 default to reg0 cause not used
    word [28:25]    = reg0;    // src2 default to reg0 cause not used
    word [23:20]    = reg0;    // dest default to reg0 cause not used
    data [127:64]  = 0;        // not used
    // nop -> op3
    word [18:15]    = 4'b0000; // nop inst op3
    word [13:10]    = reg0;    // src1 default to reg0 cause not used
    word [8:5]      = reg0;    // src2 default to reg0 cause not used
    word [3:0]      = reg0;    // dest default to reg0 cause not used
    data [63:0]    = 0;        // not used

    // one clock delay
    #halfperiod;
    #halfperiod;

    #1000 $stop;
  end

  task setreserved;
  begin
    // all these bits in the vliw word are reserved and therefore not
used.
    // they are meant for future expansion
    word [63:60] = 4'bxxxx;
    word [59] = 1'bx;
    word [54] = 1'bx;
    word [49] = 1'bx;
    word [44] = 1'bx;
    word [39] = 1'bx;
    word [34] = 1'bx;
    word [29] = 1'bx;
    word [24] = 1'bx;
```

```
      word [19] = 1'bx;
      word [14] = 1'bx;
      word [9]  = 1'bx;
      word [4]  = 1'bx;
   end
endtask

task setreset;
begin
   // do a reset
   reset = 0;
   #twocycle;
   reset = 1;
   #twocycle;
   reset = 0;
   #twocycle;
end
endtask

   vliw_top vliw_top_inst (clock, reset, word, data, readdatapipe1,
readdatapipe2, readdatapipe3, readdatavalid, jump);
   endmodule
```

Figure A.3 shows the simulation results for the register bypassing conditions of Example A.2.

Referring to Figure A.3, four register bypass conditions occur during this simulation:

1. operation1 (register r3 with contents abababababababab) to operation2 on source1 (contents int_src1datapipe2 abababababababab)

2. operation1 (register r3 with contents abababababababab) to operation2 on source2 (contents int_src2datapipe2 abababababababab)

3. operation2 (register r4 with contents 100000aaa19a8654) to operation1 on source1 (contents int_src1datapipe1 100000aaa19a8654)

4. operation3 (register r5 with contents 01111111abc739ab) to operation1 on source2 (contents int_src2datapipe1 01111111abc739ab)

Example A.3 shows the testbench for simulating a flush and jump condition during compare operation.

Example A.3 Testbench Verifying a flush and jump Condition during compare Operation

```
module vliw_top_tb();

reg clock, reset;
reg [191:0] data;
reg [63:0] word;
wire [63:0] readdatapipe1, readdatapipe2, readdatapipe3;
wire jump;

parameter halfperiod = 5;
parameter twocycle = 20;
parameter delay = 100;
```

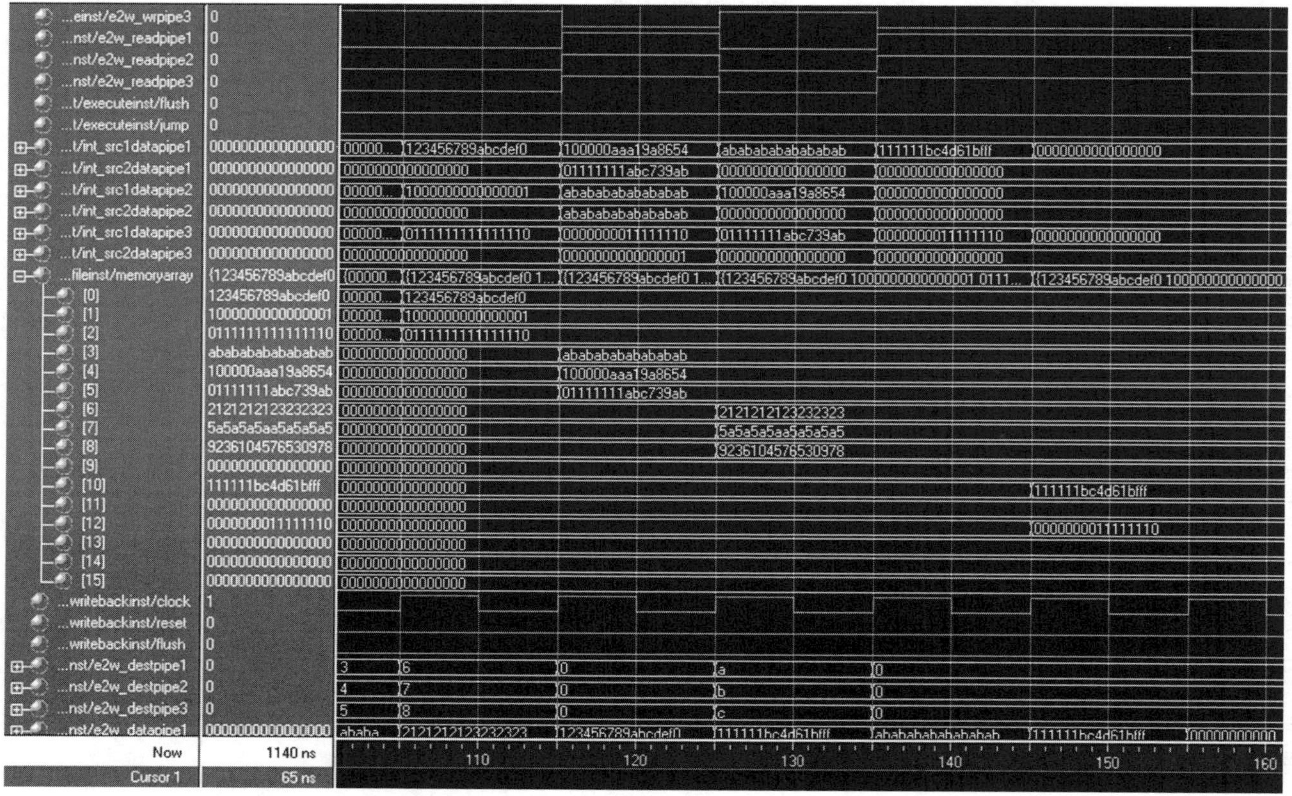

Figure A.3 Diagram showing register bypass conditions of Example A.2.

Signal	Value						
...einst/e2w_wrpipe3	0						
...nst/e2w_readpipe1	0						
...nst/e2w_readpipe2	0						
...nst/e2w_readpipe3	0						
...t/executeinst/flush	0						
...t/executeinst/jump	0						
...t/int_src1datapipe1	0000000000000000	00000...	123456789abcdef0	100000aaa19a8654	abababababababab	111111bc4d61bfff	0000000000000000
...t/int_src2datapipe1	0000000000000000	0000000000000000	01111111abc739ab	0000000000000000			
...t/int_src1datapipe2	0000000000000000	00000...	1000000000000001	abababababababab	100000aaa19a8654	0000000000000000	
...t/int_src2datapipe2	0000000000000000	0000000000000000	abababababababab	0000000000000000	0000000000000000		
...t/int_src1datapipe3	0000000000000000	00000...	0111111111111110	0000000011111110	01111111abc739ab	0000000011111110	0000000000000000
...t/int_src2datapipe3	0000000000000000	0000000000000000	0000000000000001	0000000000000000	0000000000000000		
...fileinst/memoryarray	{123456789abcdef0	{00000...	{123456789abcdef0	{123456789abcdef0 1...	{123456789abcdef0 1000000000000001 0111...	{123456789abcdef0 100000000000000000	
[0]	123456789abcdef0	00000...	123456789abcdef0				
[1]	1000000000000001	00000...	1000000000000001				
[2]	0111111111111110	00000...	0111111111111110				
[3]	abababababababab	0000000000000000	abababababababab				
[4]	100000aaa19a8654	0000000000000000	100000aaa19a8654				
[5]	01111111abc739ab	0000000000000000	01111111abc739ab				
[6]	2121212123232323	0000000000000000	2121212123232323				
[7]	5a5a5a5aa5a5a5a5	0000000000000000	5a5a5a5aa5a5a5a5				
[8]	9236104576530978	0000000000000000	9236104576530978				
[9]	0000000000000000	0000000000000000					
[10]	111111bc4d61bfff	0000000000000000				111111bc4d61bfff	
[11]	0000000000000000	0000000000000000					
[12]	0000000011111110	0000000000000000				0000000011111110	
[13]	0000000000000000	0000000000000000					
[14]	0000000000000000	0000000000000000					
[15]	0000000000000000	0000000000000000					
...writebackinst/clock	1						
...writebackinst/reset	0						
...writebackinst/flush	0						
...nst/e2w_destpipe1	0	3	6	0	a	0	
...nst/e2w_destpipe2	0	4	7	0	b	0	
...nst/e2w_destpipe3	0	5	8	0	c	0	
...nst/e2w_datapipe1	0000000000000000	ababa...	2121212123232323	123456789abcdef0	111111bc4d61bfff	abababababababab	111111bc4d61bfff	0000000000000

Now	1140 ns
Cursor 1	65 ns

```
// include the file that declares the parameter declaration for register
// names and also instruction operations
`include "/project/VLIW/64bit/simulation/regname.v"

// clock generation
initial
begin
    clock = 0;
    forever #halfperiod clock = ~clock;
end

// pump in stimulus for vliw processor
initial
begin
    // do a reset
    data = 0;
    setreserved;
    setreset;
    // word [58:55]opcode [53:50]src1 [48:45]src2 [43:40]dest op1
    // word [38:35]opcode [33:30]src1 [28:25]src2 [23:20]dest op2
    // word [18:15]opcode [13:10]src1 [8:5]src2   [3:0]dest   op3

    // load all necessary values into r0 to r8
    // load #123456789abcdef0, reg0 -> op1
    word [58:55]  = 4'b0100; // load inst op1
    word [53:50]  = reg0;    // src1 default to reg0 cause not used
    word [48:45]  = reg0;    // src2 default to reg0 cause not used
    word [43:40]  = reg0;
    data [191:128] = 64'h123456789abcdef0; // data for op1
    // load #1000000000000001, reg1 -> op2
    word [38:35]  = 4'b0100; // load inst op2
    word [33:30]  = reg0;    // src1 default to reg0 cause not used
    word [28:25]  = reg0;    // src2 default to reg0 cause not used
    word [23:20]  = reg1;
    data [127:64] = 64'h1000000000000001; // data for op2
    // load #0111111111111110, reg2 -> op3
    word [18:15]  = 4'b0100; // load inst op3
    word [13:10]  = reg0;    // src1 default to reg0 cause not used
    word [8:5]    = reg0;    // src2 default to reg0 cause not used
    word [3:0]    = reg2;
    data [63:0]   = 64'h0111111111111110;

    // one clock delay
    #halfperiod;
    #halfperiod;

    // load #abababababababab, reg3 -> op1
    word [58:55]  = 4'b0100; // load inst op1
    word [53:50]  = reg0;    // src1 default to reg0 cause not used
    word [48:45]  = reg0;    // src2 default to reg0 cause not used
    word [43:40]  = reg3;
    data [191:128] = 64'habababababababab; // data for op1
    // load #100000aaa19a8654, reg4 -> op2
    word [38:35]  = 4'b0100; // load inst op2
    word [33:30]  = reg0;    // src1 default to reg0 cause not used
    word [28:25]  = reg0;    // src2 default to reg0 cause not used
    word [23:20]  = reg4;
    data [127:64] = 64'h100000aaa19a8654; // data for op2
    // load #01111111abc739ab, reg5 -> op3
    word [18:15]  = 4'b0100; // load inst op3
    word [13:10]  = reg0;    // src1 default to reg0 cause not used
    word [8:5]    = reg0;    // src2 default to reg0 cause not used
```

```
word [3:0]    = reg5;
data [63:0]   = 64'h01111111abc739ab;

// one clock delay
#halfperiod;
#halfperiod;

// load #2121212123232323, reg6 -> op1
word [58:55]  = 4'b0100; // load inst op1
word [53:50]  = reg0;    // src1 default to reg0 cause not used
word [48:45]  = reg0;    // src2 default to reg0 cause not used
word [43:40]  = reg6;
data [191:128] = 64'h2121212123232323; // data for op1
// load #5a5a5a5aa5a5a5a5, reg7 -> op2
word [38:35]  = 4'b0100; // load inst op2
word [33:30]  = reg0;    // src1 default to reg0 cause not used
word [28:25]  = reg0;    // src2 default to reg0 cause not used
word [23:20]  = reg7;
data [127:64] = 64'h5a5a5a5aa5a5a5a5; // data for op2
// load #9236104576530978, reg8 -> op3
word [18:15]  = 4'b0100; // load inst op3
word [13:10]  = reg0;    // src1 default to reg0 cause not used
word [8:5]    = reg0;    // src2 default to reg0 cause not used
word [3:0]    = reg8;
data [63:0]   = 64'h9236104576530978;

// one clock delay
#halfperiod;
#halfperiod;

// read r0 -> op1
word [58:55]  = 4'b0110; // read inst op1
word [53:50]  = reg0;    // src1 is reg0
word [48:45]  = reg0;    // src2 default to reg0 cause not used
word [43:40]  = reg0;    // dest default to reg0 cause not used
data [191:128] = 0;      // not used
// read r1 -> op2
word [38:35]  = 4'b0110; // read inst op2
word [33:30]  = reg1;    // src1 is reg1
word [28:25]  = reg0;    // src2 default to reg0 cause not used
word [23:20]  = reg0;    // dest default to reg0 cause not used
data [127:64] = 0;       // not used
// read reg2 -> op3
word [18:15]  = 4'b0110; // read inst op3
word [13:10]  = reg2;    // src1 is reg2
word [8:5]    = reg0;    // src2 default to reg0 cause not used
word [3:0]    = reg0;    // dest default to reg0 cause not used
data [63:0]   = 0;       // not used
// one clock delay
#halfperiod;
#halfperiod;

// add r0, r1, r10 -> op1
word [58:55]  = 4'b0001; // add inst op1
word [53:50]  = reg0;    // src1 is reg0
word [48:45]  = reg1;    // src2 is reg1
word [43:40]  = reg10;   // destination is reg10
data [191:128] = 0;      // data is not used
// sub r3, r3, r11 -> op2
word [38:35]  = 4'b0010; // sub inst op2
word [33:30]  = reg3;    // src1 is reg3
word [28:25]  = reg3;    // src2 is reg3
```

```
        word [23:20]   = reg11;    // destination is reg11
        data [127:64]  = 0;        // data is not used
        // compare r4, r4, r12 -> op3
        word [18:15]   = 4'b0111;  // compare inst op3
        word [13:10]   = reg4;     // src1 is reg4
        word [8:5]     = reg4;     // src2 is reg4
        word [3:0]     = reg12;    // destination is reg12
        data [63:0]    = 0;        // data is not used

        // one clock delay
        #halfperiod;
        #halfperiod;

        // read r3 -> op1
        word [58:55]   = 4'b0110;  // read inst op1
        word [53:50]   = reg3;     // src1 is reg3
        word [48:45]   = reg0;     // src2 default to reg0 cause not used
        word [43:40]   = reg0;     // dest default to reg0 cause not used
        data [191:128] = 0;        // not used
        // read r4 -> op2
        word [38:35]   = 4'b0110;  // read inst op2
        word [33:30]   = reg4;     // src1 is reg4
        word [28:25]   = reg0;     // src2 default to reg0 cause not used
        word [23:20]   = reg0;     // dest default to reg0 cause not used
        data [127:64]  = 0;        // not used
        // read reg5 -> op3
        word [18:15]   = 4'b0110;  // read inst op3
        word [13:10]   = reg5;     // src1 is reg5
        word [8:5]     = reg0;     // src2 default to reg0 cause not used
        word [3:0]     = reg0;     // dest default to reg0 cause not used
        data [63:0]    = 0;        // not used

        // one clock delay
        #halfperiod;
        #halfperiod;

        // read r10 -> op1
        word [58:55]   = 4'b0110;  // read inst op1
        word [53:50]   = reg10;    // src1 is reg10
        word [48:45]   = reg0;     // src2 default to reg0 cause not used
        word [43:40]   = reg0;     // dest default to reg0 cause not used
        data [191:128] = 0;        // not used
        // read r11 -> op2
        word [38:35]   = 4'b0110;  // read inst op2
        word [33:30]   = reg11;    // src1 is reg11
        word [28:25]   = reg0;     // src2 default to reg0 cause not used
        word [23:20]   = reg0;     // dest default to reg0 cause not used
        data [127:64]  = 0;        // not used
        // read reg12 -> op3
        word [18:15]   = 4'b0110;  // read inst op3
        word [13:10]   = reg12;    // src1 is reg12
        word [8:5]     = reg0;     // src2 default to reg0 cause not used
        word [3:0]     = reg0;     // dest default to reg0 cause not used
        data [63:0]    = 0;        // not used

        // one clock delay
        #halfperiod;
        #halfperiod;

        // nop -> op1
        word [58:55]   = 4'b0000;  // nop inst op1
        word [53:50]   = reg0;     // src1 default to reg0 cause not used
```

```
        word [48:45]    = reg0;      // src2 default to reg0 cause not used
        word [43:40]    = reg0;      // dest default to reg0 cause not used
        data [191:128]  = 0;         // not used
        // nop  -> op2
        word [38:35]    = 4'b0000;   // nop inst op2
        word [33:30]    = reg0;      // src1 default to reg0 cause not used
        word [28:25]    = reg0;      // src2 default to reg0 cause not used
        word [23:20]    = reg0;      // dest default to reg0 cause not used
        data [127:64]   = 0;         // not used
        // nop -> op3
        word [18:15]    = 4'b0000;   // nop inst op3
        word [13:10]    = reg0;      // src1 default to reg0 cause not used
        word [8:5]      = reg0;      // src2 default to reg0 cause not used
        word [3:0]      = reg0;      // dest default to reg0 cause not used
        data [63:0]     = 0;         // not used

        // one clock delay
        #halfperiod;
        #halfperiod;

        #1000 $stop;

    end

    task setreserved;
    begin
        // all these bits in the vliw word are reserved and therefore not
used.
        // they are meant for future expansion
        word [63:60] = 4'bxxxx;
        word [59] = 1'bx;
        word [54] = 1'bx;
        word [49] = 1'bx;
        word [44] = 1'bx;
        word [39] = 1'bx;
        word [34] = 1'bx;
        word [29] = 1'bx;
        word [24] = 1'bx;
        word [19] = 1'bx;
        word [14] = 1'bx;
        word [9]  = 1'bx;
        word [4]  = 1'bx;
    end
    endtask

    task setreset;
    begin
        // do a reset
        reset = 0;
        #twocycle;
        reset = 1;
        #twocycle;
        reset = 0;
        #twocycle;
    end
    endtask

   vliw_top vliw_top_inst (clock, reset, word, data, readdatapipe1, read-
datapipe2, readdatapipe3, readdatavalid, jump);
    endmodule
```

Figure A.4 shows the simulation results of testbench shown in Example A.3.

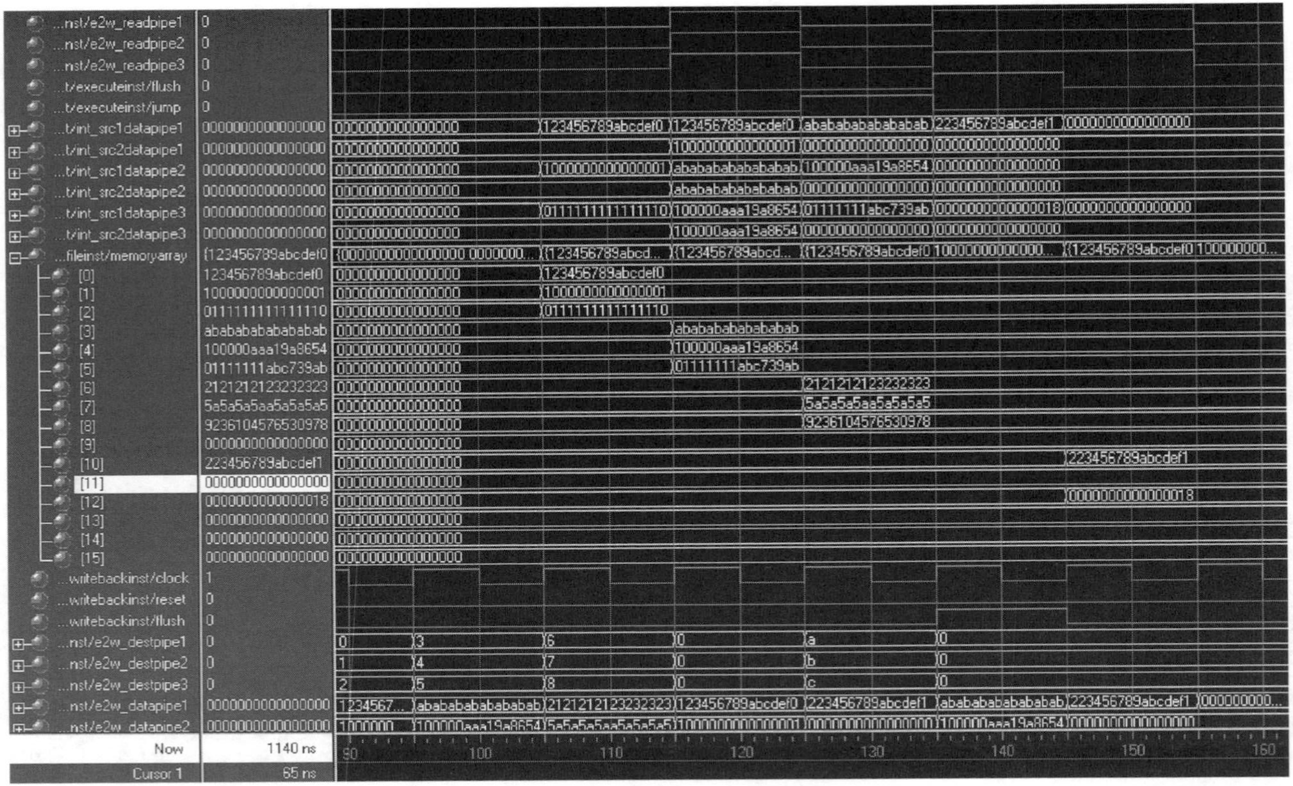

Figure A.4 Diagram showing jump and flush condition for Example A.3.

Referring to Figure A.4, the contents of register r10 are 223456789abcdef1 which are addition of contents of r0 (123456789abcdef0) and r1 (1000000000000001). The contents of register r11 are 0000000000000000 which are subtraction results of r3 (abababababababab) from r3 (abababababababab). The contents of register r12 are 0000000000000018 which are results of comparison of r4 (100000aaa19a8654) with r4. Comparison results are defined as following in Chapter 2, Section 2.1.2.

i. source1 = source2   → Branch to another instruction, a jump is required

ii. source1 > source2   → Bit 1 of destination register = 1

iii. source1 < source2   → Bit 2 of destination register = 1

iv. source1 <= source2   → Bit 3 of destination register = 1

v. source1 >= source2   → Bit 4 of destination register = 1

vi. All other bits of destination register are set to 0.

Referring to Figure A.4, when the comparison of reg r4 with r4 occurs, the signal jump goes to logic 1 indicating that a branch is taken because both source1 and source2 are equal. The signal flush goes to logic 1 one clock cycle later, allowing time for the write operation at writeback to occur. When signal flush is high, the whole VLIW microprocessor is flushed.

Synthesis Results, Gate Level Netlist

Table B.1 shows the ASIC synthesized results in terms of performance and area of the VLIW microprocessor implemented on a 0.35 micron technology.

In Table B.1, the VLIW microprocessor is limited in performance by the execute unit because it is the slowest unit due to the large ALU required to perform the VLIW microprocessor's computation. The performance for the VLIW microprocessor therefore is limited to 270 MIPS (maximum performance of 3 operations per clock cycle but varies depending on application).

Example B.1 shows a portion of the gate level netlist generated from synthesis utilizing a 0.35-micron standard cell library.

Example B.1 Verilog Gate Netlist of VLIW Microprocessor

```
//
// Verilog description for cell vliw_top,
// 07/20/05 10:41:53
//
```

TABLE B.1 Synthesis Results of VLIW Microprocessor Implemented on 0.35 Micron Technology

Module	Performance (MHz)	Transistor Count
Fetch	450	9788
Decode	600	9068
Execute	90	156374
Register file	300	147414
Writeback	900	16544

```
// LeonardoSpectrum Level 3, 2004a.30
//
module vliw_top ( clock, reset, word, data, readdatapipe1, readdatapipe2,
readdatapipe3, readdatavalid, jump ) ;

        input clock ;
        input reset ;
        input [63:0]word ;
        input [191:0]data ;
        output [63:0]readdatapipe1 ;
        output [63:0]readdatapipe2 ;
        output [63:0]readdatapipe3 ;
        output readdatavalid ;
        output jump ;

    wire r2e_src2pipe3_3_, r2e_src2pipe3_2_, r2e_src2pipe3_1_,
      r2e_src2pipe3_0_, r2e_src2pipe2_3_, r2e_src2pipe2_2_,
      r2e_src2pipe2_1_, r2e_src2pipe2_0_, r2e_src2pipe1_3_,
      r2e_src2pipe1_2_, r2e_src2pipe1_1_, r2e_src2pipe1_0_,
      ...
      ...
      nx99050, nx99051, nx99052, nx99053, nx99054, nx99055, nx99056,
      nx99057, nx99058, nx99059, nx99060, nx99061, nx99062, nx99063,
      nx99064, nx99065, nx99066, nx99067, nx99068, nx99069, nx99070,
      nx99071, nx99072, nx99073, nx99074, _9356901__XX0_XREP321,
      nx99075, nx99076, _9122375__XX0_XREP327, nx99077, nx99078,
      nx99079, nx99080, nx99081, nx99082, nx99083, nx99084, nx99085,
      nx99086, nx99087, nx99088, nx99089, nx99090, nx99091, nx99092,
      nx99093, nx99094, nx99095, nx99096, nx99097, nx99098, nx99099,
      nx99100, nx99101, nx99102, nx99103, nx99104;
      wire [951:0] \$dummy ;

    ...
    ...
    dffr reg_r2e_src2pipe3_0_ (.Q (r2e_src2pipe3[0]), .QB (\$dummy [0]),
      .D (nx39236), .CLK (clock), .R (reset)) ;
    nor02 ix39237 (.Y (nx39236), .A0 (nx95888), .A1 (nx96190)) ;
    dffr reg_r2e_src2pipe3_1_ (.Q (r2e_src2pipe3[1]), .QB (\$dummy [1]),
      .D (nx39252), .CLK (clock), .R (reset)) ;
    nor02 ix39253 (.Y (nx39252), .A0 (nx96156), .A1 (nx96190)) ;
    dffr reg_r2e_src2pipe3_2_ (.Q (r2e_src2pipe3[2]), .QB (\$dummy [2]),
      .D (nx39340), .CLK (clock), .R (reset)) ;
    nor02 ix39341 (.Y (nx39340), .A0 (nx87648), .A1 (nx96190)) ;

      ...
      ...

    inv01 ix2018 (.Y (nx2019), .A (nx2027)) ;
    inv01 ix2020 (.Y (nx2021), .A (nx803)) ;
    inv01 ix2022 (.Y (nx2023), .A (flush)) ;
    inv01 ix2024 (.Y (nx2025), .A (flush)) ;
    inv01 ix2026 (.Y (nx2027), .A (flush)) ;
endmodule
```

Table B.2 shows the FPGA synthesized results in terms of performance and area of the VLIW microprocessor implemented on an Altera Stratix EP1S25F1020C FPGA.

Referring to Table B.2, the performance for the VLIW microprocessor is limited by the execute unit to only 120 MIPS (maximum performance

TABLE B.2   Synthesis Results of VLIW Microprocessor Implemented on Altera Stratix EP1S25F1020C FPGA

Module	Performance (MHz)	Area Utilization (LC)
Fetch	230	246
Decode	250	219
Execute	40	6005
Register file	170	7172
Writeback	300	400

of 3 operations per clock cycle but varies depending on application). The FPGA implementation only has 44% performance compared to ASIC implementation.

Example B.2 shows a portion of the gate level netlist generated from the Altera Stratix EP1S25F1020C FPGA synthesis.

Example B.2   Verilog Gate Netlist of VLIW Microprocessor from FPGA Synthesis

```
//
// Verilog description for cell vliwtop_str,
// 09/21/05 17:14:35
//
// LeonardoSpectrum Level 3, 2004a.30
//
   module vliwtop_str ( clock, reset, word, data, readdatapipe1,
readdatapipe2, readdatapipe3, readdatavalid, jump ) ;

      input clock ;
      input reset ;
      input [63:0]word ;
      input [191:0]data ;
      output [63:0]readdatapipe1 ;
      output [63:0]readdatapipe2 ;
      output [63:0]readdatapipe3 ;
      output readdatavalid ;
      output jump ;

      wire flush, r2e_src2pipe3_3, r2e_src2pipe3_2, r2e_src2pipe3_1,
           r2e_src2pipe3_0, r2e_src2pipe2_3, r2e_src2pipe2_2,
           r2e_src2pipe2_1, r2e_src2pipe2_0, r2e_src2pipe1_3,
           r2e_src2pipe1_2, r2e_src2pipe1_1, r2e_src2pipe1_0,
           r2e_src1pipe3_3, r2e_src1pipe3_2, r2e_src1pipe3_1,
           r2e_src1pipe3_0, r2e_src1pipe2_3, r2e_src1pipe2_2,
           r2e_src1pipe2_1, r2e_src1pipe2_0, r2e_src1pipe1_3,
           r2e_src1pipe1_2, r2e_src1pipe1_1, r2e_src1pipe1_0,
           w2re_datapipe3_63, w2re_datapipe3_62,
           w2re_datapipe3_61, w2re_datapipe3_60, w2re_datapipe3_59,

           ...
           ...

           f2r_src1pipe1_3, f2r_src1pipe1_2, f2r_src1pipe1_1,
           f2r_src1pipe1_0, f2d_destpipe3_3, f2d_destpipe3_2,
           f2d_destpipe3_1, f2d_destpipe3_0, f2d_destpipe2_3,
           f2d_destpipe2_2, f2d_destpipe2_1, f2d_destpipe2_0,
           f2d_destpipe1_3, f2d_destpipe1_2, f2d_destpipe1_1,
           f2d_destpipe1_0, f2dr_instpipe3_3, f2dr_instpipe3_2,
```

```
                    f2dr_instpipe3_1, f2dr_instpipe3_0, f2dr_instpipe2_3,
                    f2dr_instpipe2_2, f2dr_instpipe2_1, f2dr_instpipe2_0,
                    f2dr_instpipe1_3, f2dr_instpipe1_2, f2dr_instpipe1_1,
                    f2dr_instpipe1_0, nx3476;

                    ...
                    ...


        fetch fetchinst (.word
   ({nx3476,nx3476,nx3476,nx3476,word[58],
   word[57],word[56],word[55],nx3476,word[53],word[52],word[51],word[50],
   nx3476,word[48],word[47],word[46],word[45],nx3476,word[43],word[42],
   word[41],word[40],nx3476,word[38],word[37],word[36],word[35],nx3476,
   word[33],word[32],word[31],word[30],nx3476,word[28],word[27],word[26],
   word[25],nx3476,word[23],word[22],word[21],word[20],nx3476,word[18],
   word[17],word[16],word[15],nx3476,word[13],word[12],word[11],word[10],
   nx3476,word[8],word[7],word[6],word[5],nx3476,word[3],word[2],word[1],
   word[0]}), .data ({data[191],data[190],data[189],data[188],data[187],
   data[186],data[185],data[184],data[183],data[182],data[181],data[180],

   ...
   ...


   {f2r_src1pipe2_3,f2r_src1pipe2_2,f2r_src1pipe2_1,f2r_src1pipe2_0}),
   .f2r_src1pipe3
   ({f2r_src1pipe3_3,f2r_src1pipe3_2,f2r_src1pipe3_1,f2r_src1pipe3_0}),
   .f2r_src2pipe1
   ({f2r_src2pipe1_3,f2r_src2pipe1_2,f2r_src2pipe1_1,f2r_src2pipe1_0}),
   .f2r_src2pipe2
   ({f2r_src2pipe2_3,f2r_src2pipe2_2,f2r_src2pipe2_1,f2r_src2pipe2_0}),
   .f2r_src2pipe3
   ({f2r_src2pipe3_3,f2r_src2pipe3_2,f2r_src2pipe3_1,f2r_src2pipe3_0})) ;
   decode decodeinst (.f2d_destpipe1 ({f2d_destpipe1_3,f2d_destpipe1_2,
   f2d_destpipe1_1,f2d_destpipe1_0}), .f2d_destpipe2 ({f2d_destpipe2_3,
   f2d_destpipe2_2,f2d_destpipe2_1,f2d_destpipe2_0}), .f2d_destpipe3 ({
   f2d_destpipe3_3,f2d_destpipe3_2,f2d_destpipe3_1,f2d_destpipe3_0}),
   .f2d_data
   ({f2d_data_191,f2d_data_190,f2d_data_189,f2d_data_188,f2d_data_187,
   f2d_data_186,f2d_data_185,f2d_data_184,f2d_data_183,f2d_data_182,
   f2d_data_181,f2d_data_180,f2d_data_179,f2d_data_178,f2d_data_177,
   f2d_data_176,f2d_data_175,f2d_data_174,f2d_data_173,f2d_data_172,

   ...
   ...


   d2e_datapipe3_15,d2e_datapipe3_14,d2e_datapipe3_13,d2e_datapipe3_12,
   d2e_datapipe3_11,d2e_datapipe3_10,d2e_datapipe3_9,d2e_datapipe3_8,
   d2e_datapipe3_7,d2e_datapipe3_6,d2e_datapipe3_5,d2e_datapipe3_4,
   d2e_datapipe3_3,d2e_datapipe3_2,d2e_datapipe3_1,d2e_datapipe3_0})) ;
   execute executeinst (.clock (clock), .reset (reset), .d2e_instpipe1 ({
   d2e_instpipe1_3,d2e_instpipe1_2,d2e_instpipe1_1,d2e_instpipe1_0}),
   .d2e_instpipe2
   ({d2e_instpipe2_3,d2e_instpipe2_2,d2e_instpipe2_1,d2e_instpipe2_0}),
   .d2e_instpipe3 ({d2e_instpipe3_3,d2e_instpipe3_2,d2e_instpipe3_1,
   d2e_instpipe3_0}), .d2e_destpipe1 ({d2e_destpipe1_3,d2e_destpipe1_2,
   d2e_destpipe1_1,d2e_destpipe1_0}), .d2e_destpipe2 ({d2e_destpipe2_3,

   ...
   ...


   e2w_datapipe3_3,e2w_datapipe3_2,e2w_datapipe3_1,e2w_datapipe3_0}),
   .e2w_wrpipe1 (e2w_wrpipe1), .e2w_wrpipe2 (e2w_wrpipe2), .e2w_wrpipe3
   (e2w_wrpipe3), .e2w_readpipe1 (e2w_readpipe1), .e2w_readpipe2
```

```
(e2w_readpipe2), .e2w_readpipe3 (e2w_readpipe3), .flush (flush), .jump
(jump)) ;
writeback writebackinst (.clock (clock), .reset (reset), .flush (flush),
.e2w_destpipe1
({e2w_destpipe1_3,e2w_destpipe1_2,e2w_destpipe1_1,e2w_destpipe1_0}),
.e2w_destpipe2 ({e2w_destpipe2_3,e2w_destpipe2_2,e2w_destpipe2_1,
e2w_destpipe2_0}), .e2w_destpipe3 ({e2w_destpipe3_3,
e2w_destpipe3_2,e2w_destpipe3_1,e2w_destpipe3_0}), .e2w_datapipe1 (

...
...

readdatapipe3[8],readdatapipe3[7],readdatapipe3[6],
readdatapipe3[5],readdatapipe3[4],readdatapipe3[3],
readdatapipe3[2],readdatapipe3[1],readdatapipe3[0]}),
.readdatavalid (readdatavalid)) ;
registerfile registerfileinst (.f2r_src1pipe1 ({f2r_src1pipe1_3,
f2r_src1pipe1_2,f2r_src1pipe1_1,f2r_src1pipe1_0}), .f2r_src1pipe2 (
{f2r_src1pipe2_3,f2r_src1pipe2_2,f2r_src1pipe2_1,
f2r_src1pipe2_0}), .f2r_src1pipe3 ({f2r_src1pipe3_3,
f2r_src1pipe3_2,f2r_src1pipe3_1,f2r_src1pipe3_0}), .f2r_src2pipe1 (

...
...

r2e_src2pipe2_2,r2e_src2pipe2_1,r2e_src2pipe2_0}), .r2e_src2pipe3 (
{r2e_src2pipe3_3,r2e_src2pipe3_2,r2e_src2pipe3_1,
r2e_src2pipe3_0})) ;
endmodule
module registerfile ( f2r_src1pipe1, f2r_src1pipe2, f2r_src1pipe3,
f2r_src2pipe1, f2r_src2pipe2, f2r_src2pipe3, f2dr_instpipe1,
f2dr_instpipe2, f2dr_instpipe3, clock, flush, reset,
w2re_datapipe1, w2re_datapipe2, w2re_datapipe3,
w2r_wrpipe1, w2r_wrpipe2, w2r_wrpipe3, w2re_destpipe1,
w2re_destpipe2, w2re_destpipe3, r2e_src1datapipe1,
r2e_src1datapipe2, r2e_src1datapipe3, r2e_src2datapipe1,
r2e_src2datapipe2, r2e_src2datapipe3, r2e_src1pipe1,
r2e_src1pipe2, r2e_src1pipe3, r2e_src2pipe1, r2e_src2pipe2,
r2e_src2pipe3 ) ;
input [3:0]f2r_src1pipe1 ;
input [3:0]f2r_src1pipe2 ;
input [3:0]f2r_src1pipe3 ;
input [3:0]f2r_src2pipe1 ;

...
...

output [3:0]r2e_src1pipe3 ;
output [3:0]r2e_src2pipe1 ;
output [3:0]r2e_src2pipe2 ;
output [3:0]r2e_src2pipe3 ;
wire r2e_src1datapipe1_dup0_63, r2e_src1datapipe1_dup0_62,
r2e_src1datapipe1_dup0_61, r2e_src1datapipe1_dup0_60,
r2e_src1datapipe1_dup0_59, r2e_src1datapipe1_dup0_58,

...
...

wire [128678:0] \$dummy ;
stratix_lcell reg_r2e_src2datapipe3_63 (.regout
(r2e_src2datapipe3_dup0_63), .combout (\$dummy [0]), .cout (\$dummy [1]),
.cout0 (\$dummy [2]), .cout1 (\$dummy [3]), .cin (1'b0), .dataa (
f2r_src2pipe3_int_0), .datab (nx65590), .datac (nx65591), .datad (
```

```
nx65592), .clk (clock_int), .ena (1'b1), .aclr (reset_int), .aload (
\$dummy [4]), .sclr (flush_int), .sload (\$dummy [5]), .cin0 (1'b0),
.cin1 (1'b1), .inverta (1'b0), .regcascin (1'b0), .devclrn (1'b1),
.devpor (1'b1)) ;
defparam reg_r2e_src2datapipe3_63.operation_mode = "normal";
defparam reg_r2e_src2datapipe3_63.synch_mode = "on";
defparam reg_r2e_src2datapipe3_63.lut_mask = "d080";

...
...

endmodule
```

Bibliography

Altera, http://www.altera.com

"Core Technologies: VLIW." BYTE Magazine Publication, Nov. 1994.

Fisher, Joseph A., Paolo Faraboschi, and Cliff Young. *Embedded Computing: A VLIW Approach to Architecture, Compilers and Tools*. New York: Morgan Kaufmann, 2004.

Halfhill, Tom R. "VLIW Microprocessors." ComputerWorld, Feb. 14, 2000.

Hennessey, John L., and David A. Patterson. *Computer Architecture: A Quantitative Approach, Third Edition* (The Morgan Kaufmann Series in Computer Architecture and Design). New York: Morgan Kaufmann, 2002.

The Intel 4004 Home, http://www.intel4004.com.

Lee, Weng Fook. *VHDL Coding and Logic Synthesis With Synopsys*. San Diego: Academic Press, 2000.

Lee, Wang Fook. *Verilog Coding for Logic Synthesis*. New York: John Wiley, 2003.

Patterson, David A., and John L. Hennessey. *Computer Organization and Design: The Hardware / Software Interface, Third Edition* (The Morgan Kaufmann Series in Computer Architecture and Design). New York: Morgan Kaufmann, 2004.

VLIW Research at IBM Research, http://www.research.ibm.com/vliw.

VLIW Computer Architecture, Philips Semiconductor, Pub. no. 9397-750-01759.

Xilinx, http://www.xilinx.com.

Index

A

Add (operation code), 13
Addition (arithmetic), 13
ALU (arithmetic logic unit), 2
Always block, 37–38
  flip-flops and, 38
  sensitivity list and, 37–38, 43
AND function, sensitivity list
    for, 44
And_and_nor gates, 153
And_nor gates, 152
Annotation, 169
Application-Specific Integrated
    Circuit (ASIC):
  chips of, 172
  FPGA vs., 175–176, 179
  implementing design of, 176
  structured, 178–179
  synthesis, 176
  testability and, 172–173
APR (*see* Auto place and route)
Architecture:
  of microprocessors, 3–7
  pipeline, 3–4, 20
  specifications for, 19–23
Arithmetic logic unit (ALU), 2
ASIC (*see* Application-Specific
    Integrated Circuit)
Asynchronous reset, 35–36
ATM, 2
Auto place and route (APR, block
    place and route, BPR), 164–165
  clock tree synthesis, 35
  steps of, 170
Auto/semi-custom layout, 164

B

Back annotation, 171
Back end, linking to front end,
    168–171
Barrel shift left (function), 17–18
Barrel shift left (operation
    code):
  bit format for, 18
  testbenches for verifying, 181–188
Barrel shift right (function), 18
Barrel shift right (operation
    code), 18
Basic logic gates, in standard cell
    libraries, 152
Behavioral codes, 33
BIST (built-in self test), 172
Bits, for VLIW instruction words, 18
Bitwise operators:
  logical operators vs., 39
  in RTL code, 38
Block place and route (*see* Auto place
    and route)
Blocking statements, 37–38
Borrows, in subtraction operations,
    14
Boundary scan (scan chain, JTAG
    method), 172–173
BPR (*see* Auto place and route)
Built-in self test (BIST), 172

C

Case statements:
  Complete vs. Incomplete, 42–43
  in RTL code, 39
  unwanted latch in, 39